SpringerBriefs in Environmental Science

SpringerBriefs in Environmental Science present concise summaries of cutting-edge research and practical applications across a wide spectrum of environmental fields, with fast turnaround time to publication. Featuring compact volumes of 50 to 125 pages, the series covers a range of content from professional to academic. Monographs of new material are considered for the SpringerBriefs in Environmental Science series.

Typical topics might include: a timely report of state-of-the-art analytical techniques, a bridge between new research results, as published in journal articles and a contextual literature review, a snapshot of a hot or emerging topic, an in-depth case study or technical example, a presentation of core concepts that students must understand in order to make independent contributions, best practices or protocols to be followed, a series of short case studies/debates highlighting a specific angle.

SpringerBriefs in Environmental Science allow authors to present their ideas and readers to absorb them with minimal time investment. Both solicited and unsolicited manuscripts are considered for publication.

More information about this series at http://www.springer.com/series/8868

Ratan Priya

Land Degradation in India

Linkages with Deforestation, Climate and Agriculture

Springer

Ratan Priya
Center for the Study of Regional Development
Jawaharlal Nehru University
Delhi, India

ISSN 2191-5547   ISSN 2191-5555   (electronic)
SpringerBriefs in Environmental Science
ISBN 978-3-030-68847-9   ISBN 978-3-030-68848-6   (eBook)
https://doi.org/10.1007/978-3-030-68848-6

This Springer imprint is published by the registered company Springer Nature Switzerland AG
The registered company address is: Gewerbestrasse 11, 6330 Cham, Switzerland

Contents

Chapter 1
Introduction

Abstract Land degradation is emerging as a serious threat to food security for the rising population of the world and especially for developing countries like India. This chapter has focused upon exploring a general understanding about the concept of land degradation and its status across the world with especial emphasis on South Asia. Land degradation is the temporary or permanent lowering of the productive capacity of land. Land is the fundamental means of production in an agrarian society. Without it, no agricultural production takes place, where agriculture is the mainstay of the rural Indian economy around which socio-economic privileges and deprivations revolve, and any change in its structure is likely to have a corresponding impact on the existing pattern of social inequality.

Keywords Land degradation · Wasteland · Land degradation process · South Asia · Food security · Agriculture

Degradation takes place all over the Earth at the different ways, and diverse degradation is very challenging, disputant, and significant too (Zdruli et al. 2010a, b). A great amount of the different parts of the Earth, oceans and seas, freshwater, atmosphere, and terrestrial atmosphere, has undergone the problem of degradation or moving in direction. In the following study, terrestrial degradation related to land as opposed to the sea or air has been focused only. Land degradation is insidious and spread throughout the county. Land is a serious matter of concern because livelihoods of human beings have depended on it. The uppermost surface of the Earth is always in influence of atmospheric activity, geological process, and energy radiated or transmitted by the sun. However, the amount of degradation has also stepped up by many other components, for instance, climatic factors – coldest, driest, etc. – organic factors, and most importantly uncontrolled human activities because of competition in achieving more and more development (Morgan, 2006; Mueller et al. 2014; Imeson 2012). Land is considered as a non-renewable resource and one of the most important sources for the primary production. When land from its own original condition has failed to keep or to maintain the capacity of productivity and other qualities to sustain the vegetation, then the concept of land degradation has emerged. Most significantly, it is the major source of human livelihood; thus, land degradation has become an important question that is in dispute and must be settled to fulfil the requirements of the world. The debate around the world has brought focus to land degradation as an important threat to environment. Although this problem seems to be a matter of environment, notably this is also a concern related to social and economic aspects throughout the world (Feng et al. 2005; Kiran et al. 2009a, b).

It is not surprising that most parts of food have been provided by land only; thus, it is the global issue (Southwick 1996). In 1964, two United Nations Organizations – the Food and Agriculture Organization (FAO) and International Atomic Energy Agency (IAEA) – jointly started a programme called the Soil and Water Management and Crop Nutrition Subprogramme (SWMCN). SWMCN is developed to promote nuclear-based technologies for achieving the maximum efficiency of soil, water, and nutrient in the targeted ecological zone. These techniques are helpful to reinforce the intensification of crop production and preservation of land (Nguyen et al. 2010). In 1987, the Brundtland Commission Report has focused on many burning issues of the past two decades from that time, and land degradation was also included. Land degradation leads to desertification by converting the land into unfertile land. The term "desertification" has been adopted by a UN programme to combat land degradation. Likewise, another organization, International Union of Soil Sciences (IUSS), has conducted five conferences on land degradation in Adana, Turkey (1996); Khon Kaen, Thailand (1999); Rio de Janeiro, Brazil (2001); Cartagena, Spain (2004); and Bari, Italy (September, 2008). This is not the only organization, which is dealing with land degradation; there are so many national and international organizations similar to this which have taken steps to combat the problem of land degradation (Zdruli et al. 2010a, b).

There is a big debate on land degradation at the global level. Land degradation, its extent, and its impact on agro-socio-economic condition of the people can be only adjudicated by systematically focusing on the process of land degradation and sympathizing the causes behind it. There is also a need to work on this for long term understanding of the keys to prepare the proper supervision and management of the requisite of land (Lal et al. 1997).

1.1 Land Degradation: A Concept

Land is a fundamental means of production in an agrarian society. Without it, no agricultural production takes place, where agriculture is the mainstay of the rural Indian economy around which socio-economic privileges and deprivations revolve, and any change in its structure is likely to have a corresponding impact on the existing pattern of social inequality. Land is one of the most important natural resources for surviving or feeding life. It represents and contains the quality of soil, water, flora, and fauna and involves the total ecosystem. Land degradation is the temporary or permanent lowering of the productive capacity of land (UNDP 1992). It is difficult to trace the origin of the term "land degradation", a currently widely used term that assumes diverse, overarching definitions. An international legally binding definition by the United Nations Convention to Combat Desertification (UNCCD), which has entered into force in December 1996, describes "land" as a "terrestrial bio-productive system that comprises soil, vegetation, other biota, and the ecological and hydrological processes that operate within the system", and its "degradation" as "reduction or loss ... of the biological ... productivity, ... resulting from land uses, ... or combination of ... processes, such as... soil erosion ... deterioration of ... properties of soil ... and long-term loss of natural vegetation" (UNCCD, 1994). Land degradation can be defined as lowering down the quality of land (ICAR and NAAS 2010) either in the biological productivity or usefulness of a particular place due to human interference and both (Levia 1999), hill slope (Soni and Loveson 2003) deforestation, suffering from loss of its intrinsic qualities, and decline in its capabilities (Krishan et al. 2008).

Land degradation is now an emerging danger and threat for the farming communities, spoiling or damaging agriculture and environment. Land degradation is the most severe environmental problem threatening agricultural production. Land degradation is due to rapid population growth, which results in technical backwardness in food production and lack of off-farm employment opportunities. Lindskog and Tengberg (1994) have defined land degradation as a reduction of physical, chemical, or biological status of the land, which may restrict its productive capacity. They emphasize the concept of net degradation, which is defined as a result of subtraction of natural reproduction and restorative management from natural degrading processes and human interference. Environmental degradation is a major indicator, which is restricting the growth and development of developing countries like India. Reddy (2003) has focused on land degradation in economical way in the paper and reveals that environmental degradation is defined as reduction in health cost and declining productivities of natural resources like land, water, grassland, etc. Reddy has also stated that land degradation is a very serious problem in the country at the regional level, and the loss of production has not been realized at the macro level. Desertification has occurred in some regions of India, like arid and semi-arid regions, as the resultant loss of productive cropland, dramatically increasing rangeland over a period (Ravi and Huxman, 2009).

According to Meadows and Hoffman (2003), desertification is land degradation in arid, semi-arid, and sub-humid lands. Climate change represents a key challenge to developing economies of countries like India and others too. Moreover, the nature of rainfall and the degree in changes in rainfall in coming days have also been very useful to assess the possible impact of climate on land degradation situation.

1.2 Land Degradation Process and Prospect

The degradation of productive land is a major threat to the world population as it has a direct impact on food security of the world (UNCCD 2015). The impact is intensifying due to occurring other phenomena simultaneously like climate change,

Yamuna River

IMG20201010162143.jpg
Type: JPG File
Size: 1.81 MB
Dimension: 3264 x 2448
pixels

Fig. 1.1 Grazing activities at the bank of Yamuna in Agra, Uttar Pradesh. Red colour box shows cattle grazing, which has been unplanned and can be a contributory factor for land degradation. Photo taken from the Taj Mahal Premises, Agra

biodiversity loss, and water scarcity (UNCCD Secretariat, 2013; Pulido and Bocco 2014). Land degradation is a very important geomorphic process which is occurring in many parts of the world over a range of landscape, but the causal determinant factors have local specificities which are yet to be fully understood. Like many another processes of land degradation, the gully initiation and evolution are attributed by different natural and anthropogenic causes. The natural causes of ravine erosion include climate change, catastrophic storm, tectonic uplift, etc., whereas the anthropogenic causes are deforestation, overgrazing, unplanned settlement, population pressure, and public policy (Pani and Carling 2013) (Fig. 1.1).

According to NRSA (2011), the following major categories of wastelands are identified as wasteland classes (Fig. 1.2):

Gullied/Ravine Lands: Gullies are the result of surface run-off at the local level. This surface run-off affects the unconsolidated material and results in the formation of discernible channels which cause undulating terrain. Meanwhile, ravines are the vast system of gullies developed along the river courses. These can be divided into two parts according to depth. First is the medium gullied/ravine land (depth ranging from 2.5 to 5 m), and second is deep gullied/ravine land (having depth of more than 5 m) (Fig. 1.3).

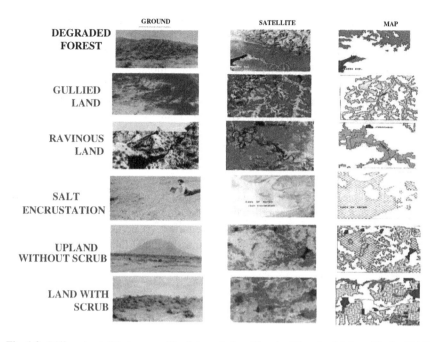

Fig. 1.2 Different satellite images of land degradation. (Source: Wasteland Atlas of India, 2011)

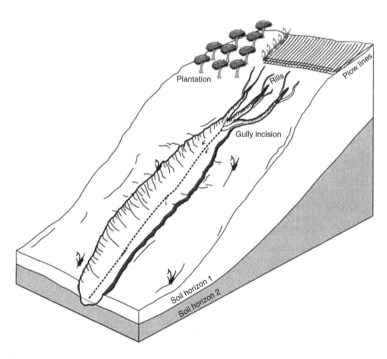

Fig. 1.3 Gully erosion process (Source: Shruthi et al. (2011). Object-based gully feature extraction using high spatial resolution imagery. Geomorphology)

Land with a Scrub: This covers the areas that are dominated by shrubs and possesses shallow and *skeleton* soil, chemically degraded with extreme slopes, severe erosion, and excessive aridity. They are intermixed with the cropped areas. The colour of these areas depends on the surface moisture ranging from yellow to brown to greenish blue. They also vary in size from small to large and are either contiguous or in a dispersed pattern. Such land can be divided into two parts: (I) dense, having vegetation cover more than 15% with moderate slope in foothills and plains and surrounded by the agricultural land, and (II) open, having sparse vegetation cover less than 15% with thin soil, and it is also prone to degradation due to erosion.

Waterlogged/Marshy Land: Waterlogged land is that land where water stands (at/near the surface) for most of the year. Marshland is that land which is inundated by water (permanently or periodically) and covered by vegetation. Depending on duration, this can be divided into two categories: (i) permanent waterlogged/marshy land and (ii) seasonal waterlogged/marshy land (Table 1.1).

Table 1.1 Indicators of land degradation

S. No.	Category	Indicators	Substantiation
1.	Agriculture	Abandoned land (Doucha and Vanek 2006)	Land on which agriculture is not being done now (proportion to total agricultural land)
		Arable land (Doucha and Vanek 2006)	Land which is available for cultivation (proportion to total agricultural land)
		No. of cattle (Doucha and Vanek 2006)	Cattles per farmer
		Shifting cultivation (NRSA 2011)	Constitute current jhum, abandoned jhum, underutilized/degraded notified forest land, agricultural land inside notified forest land, degraded pasture/grazing land, and degraded land under plantation crop
2.	Economic	Mining/industrial wasteland (NRSA 2011)	Industrial wasteland, mining wasteland
3.	Geographic	Sand (desert/coastal/ravine) (NRSA 2011)	Coastal sand, riverine sand, desertic sand (dunes less than 15 m high and interdunal areas), desertic sand – moderately high dunes and high dunes
		Barren/rocky/stony waste (NRSA 2011)	
		Snow covered/glacial area (NRSA 2011)	

Source: Table prepared by the author by using various sources mentioned in the citation

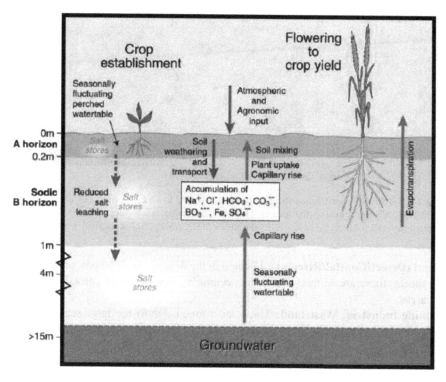

Fig. 1.4 Soil salinity process. (Source: Vermang et al. (2011). Land degradation processes and assessment)

Land Affected by Salinity and Alkalinity: Such type of land has an adverse effect on the growth of plants due to the presence of the excess soluble salts or sodium. Considering the extent of salinity in the soil, it can be divided into two subdivisions: (i) land affected by salinity/alkalinity (strong), Land affected by salinity and alkalinity: Strong: electrical conductivity (EC) levels (dS/m) more than 30, pH more than 9.8, and exchangeable sodium percentage (ESP) more than 40, and (ii) land affected by salinity/alkalinity (medium), electrical conductivity (EC) levels (dS/m) between 8 and 30, pH between 9.0 and 9.8, and exchangeable sodium percentage (ESP) between 15 and 40 (Fig. 1.4 and Fig. 1.5).

Shifting Cultivation: A traditional agricultural practice of growing crops on forested and vegetated hillslopes by using "slash and burn" method. It is known by different names in different regions, such as jhumming in northeastern region of India and Podu cultivation in the southern states of India, etc. This type of cultivation is mostly associated with mountainous or the hilly areas midst forest cover and forest cleared areas. It can be divided into two subdivisions, (i) current jhum and (ii) abandoned jhum.

Fig. 1.5 [A] Salty "Usar", which also tastes salty (according to a local farmer). [B] Big "Kankar" formation. [C] Red circle is the area of "Kankar" formation below the top layer of the soil. (Photograph taken during the fieldwork by the author in December 2017 at Kannauj, Uttar Pradesh)

Sand (Desert/Coastal/Riverine): Located in the deserts, on riverbeds, or along the shores, these are formed due to the accumulation of sands in coastal or inland areas.

Mining/Industrial Wasteland: These are formed due to the large-scale mining operations which resulted in the degradation of land and mine dumps.

1.3   Land Degradation Status in the World

Focusing on the land degradation in developing countries, the paper of Kishk (1990) inferred that in developing countries, people are growing towards the awareness of expansion of cultivated areas to fulfil the basic needs of a fast-growing population. In developing countries, agriculture is more important in economy. Increasing pressure of population forces people to extract more and more from the land because of scarcity of cultivated areas. Poor farmers are not able to conserve their land due to misuse and poor management. The main affecting factor for land degradation is man, although some environmental processes are also responsible, but this environmental problem is the result of human activities. Demographic processes influence land degradation through the intervening variables of land use. Economic variables related to different anthropogenic activities force to induce changes in land use pattern of the region (Bilsborrow and Ogendo 1992). Rapid land degradation in developing countries has happened particularly due to soil erosion in sloping uplands. Changing the land use practices in place of current land use practices, which have made undesirable changes of land and forest degradation, should be of concern. Who is responsible for this? The answer is not only farmers because thay are forced to invest low in sustainale practices given their poor economic status. These poor people require the help from the government, which must be made accountale first. Availability of appropriate technologies is required to deal with land degradation and conservation in developing countries. Right approach towards such problems, proper thinking, and taking into consideration all relevant socio-economic and cultural aspects must be good step in mitigating most of the trouble associated with land degradation. No theories can explain the complex inter-action between population growth and changes in land use. Understanding the causes of environmental degradation in rural areas in developing countries and the factors that lead to the changes in land use is useful to mark the reason for land degradation in developing countries. Productivity changes in lowland food agricul-ture could heavily reduce the rate of land degradation in upland by converting the relative profitability of food and tree crops in favour of the latter.

If we look at the different issues on problems related to land degradation in developing countries, Kishk (1986) has deduced that the Egyptian population has grown very rapidly in recent years, and despite the efforts made towards industrial-ization, the major needs are met from agriculture. The risk of desertification is very high in Egypt. Most of the areas are subject to sand movement, soil stripping, salin-ity, and human pressure. Land degradation causes destruction and irreversible losses to the highly productive cultivated land in Nile valley and deltas. Industrial produc-tion of bricks and other building materials, extracted from alluvial fertile soil, causes the removal of the topsoil. Losses in Egyptian economy have occurred due to land degradation. According to Barker and McGregor (1988), the Yallahs Basin of Jamaica is only 180 km², but it has a high population density and suffers from land degradation and rural poverty. The Yallahs Basin has a high natural propensity for erosive activity due to the presence of geologically recent mountain range, which is topographically defined by steep slopes and dissected terrain with sharp ridges and deep gullies. Putting the words on land degradation in Pakistan, Ellis et al. (1993) have stated that deforestation occurred in Pakistan and it recorded clearing of the original woodland at large scale. Due to the removal of vegetation, soil comes directly exposed to rainfall. A very intense rainfall causes serious soil erosion. Vegetation of Murree Hills (Pakistan) has been totally cleared, which was associ-ated with soil erosion and environmental problem. The layer of the topsoil is becom-ing thinner. Thus, it is difficult to cultivate and reforest and creates high run-off responses, which results in shifting of farmland and flooding in the adjacent low-land areas.

A study by Stahl (1993) has explained the condition of land degradation in East Africa. East Africa has an annual population growth of 3% like rest of the continent. It is in the mid-stage of a demographic transition period of high fertility and mortal-ity rate. Land degradation is threatening the very basis of East African farmer's society. Although land degradation is the loss of such a precious thing, which could not be returned back, there are a number of steps taken by the government to miti-gate or prevent further degradation. If we talk about China, in Jilin Province, the dominant natural vegetation type in the region is meadow steppe, characterized by drought and alkaline tolerant grass species with the loess land. Agriculture in West Jilin has developed very fast in the past decades. However, land degradation is a serious problem in West Jilin; this is one of the most important areas for commercial production. Land degradation is identified, when characteristics such as increasing desertification and soil fertility decline have been found in this region. This paper has calculated the land quality based on the depth of soil humus, net income, and farmer's evaluation of land quality changes. Four qualities have been found in one zone around labour centres: enhancing, conserving, degrading, and exhausting. This paper inferred that agricultural intensification is associated with land improvement near the field, but it causes land degradation in the remote field.

Mambo and Archer (2007) try to examine how political, economic, and environmental factors influenced the status of land degradation in Zimbabwe. The author studies the land degradation map in agro-ecological zones. Very high probability of land degradation is found in extensive farming and regarded as arid areas. High susceptibility of degradation is found in semi-intensive farming zone under woodland in vegetation. Moderate susceptibility occurred in arid areas with any vegetation including bush land and cultivation. Low probability of land degradation exists in semi-arid eras with any vegetation and land class type. Land degradation is an issue of global environment. Most countries are suffering in terms of the amount of degradation and economic impact. Land degradation is a long-term decline in ecosystem function and productivity.

1.4   Land Degradation Status in South Asia

Regarding and concerning land degradation in SAARC countries, 83 million hectares or 25% of total areas under crops and pasture are affected by water erosion in the South Asian region. Salinity and water logging are major factors for land degradation in irrigated areas and in coastal areas. Salinization mainly occurs in the Indus river basin, southern coastline of Sri Lanka. Loss of nutrient and depletion of organic matter are another form of degradation. Nutrient depletion occurs in mid-altitude hills of Nepal, Northern India. Water erosion spread over the foothills of Himalayas, and river bank erosion occurs majorly in the floodplain of Ganges, Brahmaputra, Yamuna, Teesta, and Meghna Rivers. Wind-eroded areas are Western Rajasthan, the coastal region of India, and the dry region of Pakistan. Desertification is land degradation being present in the dryland of India and Pakistan. This area mostly suffers from moisture stress, sand movement, high wind velocity, and very limited canopy cover. Agrochemical pollution has been found in Pakistan due to heavy use of agrochemicals. By introducing green revolution, undoubtedly, food productivity increased many times. However, unsustainable packaging practices, high population pressure, and uncertainty in monsoon have added to various kinds of land degradation (Singh and Singh 2011).

The rapid population growth and concurrent increase in the demand for agriculture, food, water, and shelter in Bangladesh have put more pressure on land and water resources. In Bangladesh, approximately 220 ha (1% of total cultivated areas) of land goes out cultivated per day. This has serious consequences on the sustainability of agricultural development potential, food supply, and food security of the country. Soil erosion in Bangladesh was found in the parts of level to gently undulating high terraces of Madhupur, Barind, and Akhaura tracts, in terms of top-soil and nutrient loss. Water erosion is a serious problem in Bangladesh, because of high seasonal rainfall, low organic matter content, poor soil structure and management, and rapid destruction of vegetation cover in different slope of hills, continuously thinning the layer of the topsoil. Water erosion covers all types of soil erosion like sheet and rill and gully erosion. Wind erosion and salinization are also major problem in the country. Approximately 1.7, 3.2, and 3.1 million ha area has degraded due to water erosion, soil fertility decline, and salinization, respectively. Plantation, organic agriculture, preserving soil fertility, fertilizer management, active participation of people, and mangrove plantation in coastal areas are some suggestions to reduce land degradation.

Land degradation in Bhutan is a natural phenomenon as well as man-made. In the dynamic mountain setting of Bhutan, land degradation is a natural and inevitable process. Land degradation is dominated by water as a degrading agent through surface erosion (splash, sheet, and rill), gully formation, bank erosion, and (flash) flooding. Mass movement driven by gravity is a secondary but often very destructive process, often interacting closely with water-induced degradation. Mass movement, flood, and soil erosion driven by gravity, water flow on steep mountain slopes has been outcome of complex geological and geomorphological process. In Bhutan, land degradation results in increased sediment transport by mountain streams. Unsustainable agriculture, forest degradation, forest harvesting, forest fires, livestock rearing and grazing, land use intensification and competition, mining and quarrying, infrastructure development, and policy gap are factors which are responsible for land degradation in Bhutan. Institutional setting for land and environmental management, with the National Environment Commission (NEC) and the National Land Commission (NLC) are the highest decision-making bodies, is important to mitigate land degradation. Sustainable land management project, SLM planning tools, National Action Program (NAP), and Land Management Campaign are some programs adopted to mitigate land degradation (Dorji 2011).

Land degradation is a serious phenomenon in Nepal, because Nepalese economy is based on agriculture with 70% of population engaged in it. The main cause of land degradation in Nepal is the poor economic condition, lack of knowledge and awareness, and inefficient government policies. The terrain, soil, geology, climate, and land management practices are very complex, so it is difficult to measure land degradation in Nepal (Khadka and Sharma 2011).

Concerning deforestation in India, a paper of Menon and Bawa (1998) identifies that there are two serious effects of deforestation on environmental degradation and loss of biodiversity. This also extremely emphasizes that deforestation in India must be of special concern because of the diversity of the subcontinent. Regarding deforestation or transformation in forest ecosystem (Singh et al. 2007), soil biological, biochemical, and microbiological changes are very sensitive to make changes in soil conditions, which is due to forest degradation, and this directly affects the ecosystem stability and fertility. This is directly linked with land degradation because land degradation is also directly linked with reduction in land productivity, because of human activities like deforestation. Thus, the degradation of forest ecosystem by anthropogenic factor causes nitrogen transformation rate; therefore, deforestation may be one of the factors affecting the productivity of land. Kiran et al. (2009a, b) give a frame of land degradation in utilization context, because an original and undisturbed land would never be degraded automatically, which means the main cause is human activity. In utilization context, land degradation means loss of biological agricultural productivity or erosion in the land's capacity to support desirable vegetation and to maintain the yield level over the years of use. This paper has accounted that India is having 8% of the biodiversity of the world despite having only 2.4% of total land. Nevertheless, another story is that a very large part of India has degraded land as well as continually reduced medicinal plant species, which results in loss of diversity and moving towards dying. The solution for all these problems is cultivation of medical plant on degraded land, which is helpful to maintain our diversity, minimize pressure on crop, increase the sources of good income because of demand of medical plant in the world market, and reclaim the degraded land in the country. Thus, deforestation might be considered as one of the factors that affect the land degradation, that is why it is important to study the forested land in the study area.

Regarding land degradation in India, Venkateswarlu and Prasad (2011) inferred that land degradation is steadily increasing due to the growing pressure on land and unsustainable land use in India. In India, 55.27 million ha of land is under degraded land, contributed by gullied land (1.19 M ha), land with or without scrub (18.80 M ha), waterlogged (0.97 M ha), saline/alkali (1.20 M ha), shifting cultivation (1.88 M ha), degraded forest and agriculture land under forest (12.66 M ha), degraded pasture, sand, mining and industrial wastelands, and barren/stony/snow cover. Soil degradation increases with increased use of land (Pohit 2013). The causes of land degradation in India are water erosion, wind erosion, waterlogging, salinization, acidification, soil physical constraints, flood and droughts, vegetation, nutrient mining, depletion of soil organic matter, over-exploitation of groundwater, and use of poor-quality groundwater (Sharda 2011). Acid soils constitute about 30% of the total cultivated area in India. The productivity of acid soil is low due to low pH; presence of toxic levels of Al, Fe, and Mn; nutrient imbalance; deficiency of Ca, Mg, S, P, B, and Mo; and poor microbial activity (Jena 2011). Land degradation has also occurred due to selenium (se). Selenium is not only an essential element for plant growth, but also its concentration in plant tissues is important for animal and human health. Approximately 1000 ha of Se-degraded land has been characterized and mapped in northwestern India.

Linking with land degradation in the different part of India, Gupta et al. (1998) tell in this paper that water erosion and mining are the main causes of land degradation in Palamau district, Nagpur Plateau. The degraded lands in the district should be rehabilitated with due regard to long-term sustainability. Severely to very severely eroded areas can be taken care of by adopting certain measures like field bounding and levelling, contour cultivation, dry farming, safe disposal of excess rainwater, strip cropping, and following proper crop rotation. Gullies can be stabilized both by vegetative and mechanical measures. NCT Delhi has significant variation in its soil characteristics and agricultural capability: deep and fertile land in Haryana and Uttar Pradesh region, plateaus and midlands of Rajasthan region, and ridge and rocky outcrops in part of Haryana and NCT Delhi. The region extending over an area of 30,242 km^2 is undergoing unprecedented growth, causing serious ecological degradation, conflicting land use practices, and conversion of agricultural landmass for other uses (Paliwal et al. 1999). Thar Desert is the most densely populated desert region of the world. Increase in population and decrease in common property resources have resulted in increased pressure on scarce water resources in these water-limited landscape. Thar Desert is affected by rapid soil degradation and vegetation loss (Ravi and Huxman 2009). The main problems detected in the Himalayan region are erosion, wetland loss, and the vegetation change. Erosion assessment reveals that more than 48.27% of the area in the valley is under very high erosion risk. The paper found that the Pir Panjal watersheds are under high erosion risk as compared to the Greater Himalayan watersheds due to weak lithological formation. The Karewa Formation is mainly found to be under very sparse vegetation cover and is extensively used for soil excavation for construction and horticultural purposes. The Karewa Formation found more widespread along with the Pir Panjal range is tectonically very active. It was found that the Pohru and Doodhganga watersheds of the Pir Panjal range are under very high erosion (Zaz and Romshoo 2012).

Though India is an agrarian economy, nevertheless economic activities relate to land (Reddy 2003; Khoshoo et al. 2009) in different ways. Literatures have emphasized that land in India is experiencing different types of degradation (Pani and Carling 2013) due to unstable and unplanned practices (Maconachie 2007; Morgan 2006; Natarajan et al. 2010; Isaac et al. 2010). India has a vast variation in temperature and rainfall under monsoon effects. Climatic factors and phenomenon are also affecting the pace of land degradation (Hong and Ju 2007; Clarke and Rendell 2007). Human activities like deforestation, overgrazing, mining, and quarrying are factors affecting land degradation too (Sahu and Dash 2011; Turnbull et al. 2014). Human activities have not only brought about degradation of land but also cause to aggravate the pace of natural factors to force damage to land (Mortimore 1993; Jolly 1993; Lal 2009). Soil is the most important renewable resource. Loss of nutrients in soil is recorded due to loss of the topsoil layer (Bojo 1991; Pani et al. 2011; Braun and Gerber 2012). Because of the given reason, the agricultural productivity is lying down in India (National Commission on Agriculture 1976; Gupta et al. 1998; Bhattacharya and Guleria 2012; Zaz and Romshoo 2012). At present, there are about130 million ha of degraded land in India. Approximately, 28% of it belongs to the category of forest-degraded area, 56% of it belongs to water-eroded area, and the rest is affected by saline and alkaline activities.

References

Bojo, J. P. (1991). Economics and Land degradation. Ambio, 20 (2), 75–79. Retrieved from http://www.jstor.org/stable/4313780

Barker, D. & McGregor, F. M. (1988). Land degradation in Yallahs Basin, Jamaica: Historical Notes and Contemporary Observations. Geography, 73 (2), 116–124. Retrieved from http://www.jstor.org/stable/40571382

Bhattacharya, T & Guleria, S. (2012). Costal Flood Management in Rural Planning Unit Through Land-Use Planning: West Bengal, India. Journal of Coastal Conservation, 16, 77–87. https://doi.org/10.1007/s11852-011-0176-x.

Braun, J. V. & Gerber, N. (2012). The Economics of Land and Soil Degradation- Toward an Assessment of the Cost of Inaction. In R. Lal et al. (Eds.), Recarbanization of the Biosphere: Ecosystems and the Global Carbon Cycle, (pp.493–516). https://doi.org/10.1007/978-94-007-4159-1_23

Bilsborrow, R. E. & Ogendo, H. W. O. O. (1992). Population-Driven Changes in Land Use in Developing Countries. Ambio, 21 (1), 37–45. Retrieved from http://www.jstor.org/stable/4313884

Daucha, T. & Vinek, D. (2006). Interactions between Agricultural Policy and Multifunctionality of Czech Agriculture. Chapter in book, Coherence of Agricultural and Rural Development Policies. ISBN-9264023887

Dorji, K. D. (2011). Strategies for Arresting Land Degradation in Bhutan. In Dipak Sarkar, A. B. Azad, S. K. Singh & N. Akter (Eds.), Strategy for Arresting Land Degradation in South Asian Countries, (pp. 239–250). OECD.

Ellis, S., Taylor, D. & Masood, K. R. (1993). Land Degradation in Northern Pakistan. Geography, 78 (1), 84–87. Retrieved from http://www.jstor.org/stable/40572233

Feng, J., Wang, T., Qi, S. & Xie, C. (2005). Land Degradation in the Source Region of the Yellow River, Northeast Qinghai-Xizang Plateau: Classification and Evaluation. Environmental Geology, 47, 459–466. https://doi.org/10.1007/s00254-004-1161-6

Jena, D. (2011). Acid Soil Management in India-Challenges and Opportunities. In Dipak Sarkar, A. B. Azad, S. K. Singh & N. Akter (Eds.), Strategy for Arresting Land Degradation in South Asian Countries, (pp. 172–190). Dhaka, Bangladesh, SAARC Agriculture Centre.

Jolly, C. L. (1993). Population Change, Land Use and the Environment. Reproductive Health Matters, 1, 13–25. Retrieved from http://www.jstor.org/stable/3774852

Gupta, S. K., Ahmed, M., Hussain, M., Pandey, A. S., Singh, P., Saini, K. M, & Das, S.N. (1998). Inventory of Degraded Lands of Palamau District, Bihar- A Remote Sensing Approach. Journal of the Indian Society of Remote Sensing, 26 (4), 161–168.

Hong, M. & Ju, H. (2007). Status and Trends in Land degradation in Asia. In Mannava V. K. Sivakumar & Ndegwa Ndiang'ui (Eds.), Climate and Land Degradation, (pp.55–64). NY, Springer Berlin Heidelberg.

Clarke, M. L. & Rendell, H. M. (2007). Climate, Extreme Event and Land Degradation. In Mannava V. K. Sivakumar & Ndegwa Ndiang'ui (Eds.), Climate and Land Degradation, (pp.137–149). NY, Springer Berlin Heidelberg.

Khadka, Y. G. & Sharma, P. (2011). Land Degradation and Rehabilitation in Nepal. In Dipak Sarkar, A. B. Azad, S. K. Singh & N. Akter (Eds.), Strategy for Arresting Land Degradation in South Asian Countries, (pp. 133–149). Dhaka, Bangladesh, SAARC Agriculture Centre.

Khoshoo, T.N. & Tejwani, K.G. (1993). Soil erosion and conservation in India (status and policies). In Pimentel, D. (Ed.), World Soil Erosion and Conservation. pp. 109–146. Cambridge: Cambridge University Press.

Kishk, M. A. (1986). Land Degradation in the Nile valley. Ambio, 15 (4), 226–230. Retrieved from http://www.jstor.org/stable/4313255

Kishk, M. A. (1990). Conceptual Issues in Dealing with Land Degradation/Conservation Problems in Developing Countries. GeoJournal, 20 (3), 187–190. Retrieved from http://www.jstor.org/stable/41144633

Kiran, Kudesia, R., Rani, M. & Pal, A. (2009a). Reclaiming Degraded Land in India Through the Cultivation of Medical Plants. Botany Research International, 2 (3), 174–181.

Kiran, Pal, A., & Kudesia, R. (2009b). Soil Quality of Degraded Land of Bundelkhand region with Special Reference to Jhansi districts of Uttar Pradesh. Journal of Phytology, 1 (5). 328–332. URL- http://journal-phytology.com/index.php/phyto/article/view/771/634

Krishan, G., Kushwaha, S.P.S., & Velmurugan, A. (2008). Land Degradation Mapping in the Upper Catchment of River Tons. Journal of Indian Society of Remote Sensing, 37, 119–128.

Lal, R. (2009). Soil Degradation as a Reason for Inadequate Human Nutrition. Food Security, 1, 45–57. https://doi.org/10.1007/s12571-009-0009-z.

Lal, R., Wagner, A., Greenland, D. J., Quine, T., Billing, D. W., Evans, R. & Giller, K. (1997). Degradation and Resilience of Soils [and Discussion]. Philosophical Transactions: Biological Sciences, 352 (1356), 997–1010. Retrieved from http://www.jstor.org/stable/56542

Levia, D. F. (1999). Land Degradation: Why Is It Continuing? Ambio, 28 (2), 200–201. Retrieved from http://www.jstor.org/stable/4314877

Lindskog, P. & Tengberg, A. (1994). Land Degradation, Natural Resources and Local Knowledge in the Sahel Zone of Burkina Faso. Geo Journal, 33 (4), 365–375. Retrieved from http://www.jstor.org/stable/41146235

Imeson, A. (2012). Desertification, land degradation and sustainability. UK, John Wiley & Sons, Inc.

ICAR & NAAS. (2010). Degraded and Wastelands of India Status and Spatial Distribution. Published by Dr T P Trivedi, Project Director, Directorate of Information and Publications of Agriculture, Indian Council of Agricultural Research, Krishi Anusandhan Bhavan I, Pusa, New Delhi. URL- http://www.icar.org.in/files/Degraded-and-Wastelands.pdf

Isaac, R. K., Sharama, D. P. & Swaroop, N. (2010). New Approaches in Reclamation of Degraded Soils with Special Reference to Sodic Soil: An Indian Experience. In P. Zdruli, M. Pagliai, S. Kapur, & A. F. Cano (Eds.), Land Degradation and Desertification: Assessment, Mitigation and Remediation, (pp. 253–266). London, NY, Springer Dordrecht Heidelberg.

Mortimore, M. (1993). Population Growth and Land Degradation. GeoJournal, 31 (1), 15–21. Retrieved from http://www.jstor.org/stable/41145902

Maconachie, R. (2007). Urban Growth and Land Degradation in Developing Cities: Changes and Challenges in Kano Nigeria. Hampshire, England: Ashgate Publication Limited.

Mambo, J. & Archer, E. (2007). An assessment of Land Degradation in the Save Catchment of Zimbabwe. Area, 39 (3) 380–391. Retrieved from http://www.jstor.org/stable/40346053

Meadows, M. E. & Hoffman, T. M. (2003). Land Degradation and Climate Change in South Africa. The Geographical Journal, 169, 168–167. Retrieved from http://www.jstor.org/stable/3451397

Menon, S. & Bawa, K. S. (1998). Deforestation in the Tropics: Reconciling Disparities in Estimates for India. Amibio, 27 (7), 576–577. Retrieved from http://www.jstor.org/stable/4314794

Morgan, R. P. C. (2006). Managing Sediment in the Landscape: Current Practices and Future Vision. In P. N. Owens (Eds.), Soil Erosion and Sediment Redistribution in River Catchments, (pp.287–293). Oxfordshire, UK, Biddles Ltd, King's Lynn.

Mueller, E. N., Wainwright, J., Parsons, A. J. & Turnbull, L. (2014). Land degradation in Drylands: An Ecogeomorphological Approach. In E. N. Mueller, J. Wainwright, A. J. Parsons & L. Turnbull (Eds.), Patterns of Land Degradation in Drylands: Understanding Self-Organised Ecogeomorphic Systems, (pp. 1–12). London, NY, Springer Berlin Heidelberg.

Natarajan, A., Janakiraman, M., Manoharan, S., Kumar, K. S. A., Vadivelu, S. & Sarkar, D. (2010). Assessment of Land Degradation and Its Impacts on Land Resources of Sivagangai Block, Tamil Nadu, India. In P. Zdruli, M. Pagliai, S. Kapur, & A. F. Cano (Eds.), Land Degradation and Desertification: Assessment, Mitigation and Remediation, (pp. 235–252). London, NY, Springer Dordrecht Heidelberg.

National Commission on Agriculture (1976). Report. Ministry of Agriculture & Irrigation, Government of India, New Delhi.

Nguyen, M., Zapata, F. & Dercon, G. (2010). "Zero-Tolerance" on Land Degradation for Sustainable Intensification of Agricultural Production. In P. Zdruli, M. Pagliai, S. Kapur, & A. F. Cano (Eds.), Land Degradation and Desertification: Assessment, Mitigation and Remediation, (pp. 37–48). London, NY, Springer Dordrecht Heidelberg.

NRSA & Department of Land Resources. (2011). District and Category-Wise Wastelands of India. Ministry of Rural Development, New Delhi, GOI & Indian Space Research Organization, Hyderabad. URL- http://www.dolr.nic.in/WastelandsAtlas2011/Wastelands_Atlas_2011.pdf

Paliwal, R., Geevarghese, G. A., Babu, P. R. & Khanna, P. (1999). Valuation of Landmass Degradation Using Fuzzy Hedonic Method: A Case Study of National Capital Region. Environment and Resources Economics, 14, 519–543.

Pani, P. & Carling, P. (2013). Land Degradation and Spatial Vulnerabilities: A Study of Inter-Village Differences in Chambal Valley, India. Asian Geographer, 30 (1), 65–79. Retrieved from https://doi.org/10.1080/10225706.2012.754775

Pani, P., Mishra, D. K. & Mohapatra, S. N. (2011). Land Degradation and Livelihoods in Semiarid India: A Study of Formers' Perception in Chambal Valley. Asian Profile, 39 (5), 505–519.

Pohit, S. (2013). Land Degradation and Trade Liberalization: An Indian Perspective. MPRA, NISTADS-CSIR, 1–22. Retrieved from http://mpra.ub.uni-muenchen.de/44496/

Pulido, J. and Bocco, G. (2014). Local Perception of Land degradation in Developing Countries: A Simplified Analytical Frameworkof Driving Forces. Processes, Indicators and Coping Strategies. Living Reviews Landscape Research. 8 (4). https://doi.org/10.12942/lrlr-2014-4

Ratna Reddy, V. (2003). Land Degradation in India: Extent, Costs and Determinants. Economic and political weekly 38(44):4700–4713. https://doi.org/10.2307/4414225

Ravi, S. & Huxman, T. E. (2009). Land Degradation in Thar Desert. Frontier in Ecology and Environment, 7 (10), 517–518. Retrieved from http://www.jstor.org/stable/25595242

Sharda, V. N. (2011). Strategies for Arresting Land Degradation in India. In Dipak Sarkar, A. B. Azad, S. K. Singh & N. Akter (Eds.), Strategy for Arresting Land Degradation in South Asian Countries, (pp. 75–132). Dhaka, Bangladesh, SAARC Agriculture Centre.

Sahu, H. B. & Dash, S. (2011). Land Degradation due to Mining in India and its Mitigation Measures. 2nd International Conference on Environmental Science and technology, IPCBEE, 6, 132–136. Singapore, IACSIT Press.

Shruthi, R. B. V., Kerle, N., & Jetten, V. (2011). Object-based gully feature extraction using high spatial resolution imagery. Geomorphology, 134(3–4), 260–268. https://doi.org/10.1016/j.geomorph.2011.07.003

Singh, A. K. & Singh, S. K. (2011). Strategies for Arresting Land Degradation in South Asian Countries. In Dipak Sarkar, A. B. Azad, S. K. Singh & N. Akter (Eds.), Strategy for Arresting Land Degradation in South Asian Countries, (pp. 1–31). Dhaka, Bangladesh, SAARC Agriculture Centre.

Singh, R. S., Tripathi, N. & Singh S. K. (2007). Impact of Degradation on Nitrogen Transformation in a Forest Ecosystem of India. Environmental monitoring and Assessment, 125, 165–173.

Stahl, M. (1993). Land Degradation in East Africa. Ambio, 22 (8), 505–508. Retrieved from http://www.jstor.org/stable/4314139

Soni, A.K. & Loveson, V.J. (2003). Land Damage Assessment: A Case Study. Journal of Indian Society of Remote Sensing, 31 (3), 175–186. https://doi.org/10.1007/BF03030824

Southwick, C. H. (1996). Global Ecology in Human Perspective. New York, Oxford University Press, Inc.

Turnbull, L., Wainwright, J. & Ravi, S. (2014). Vegetation Change in the Southwestern USA: Pattern and Process. In E. N. Mueller, J. Wainwright, A. J. Parsons & L. Turnbull (Eds.), Patterns of Land Degradation in Drylands: Understanding Self-Organised Ecogeomorphic Systems, (pp.289–314). London, NY, Springer Berlin Heidelberg.

UNCCD (1994). United Nations Convention to Combat Desertification in Countries Experiencing serious drought and/or Desertification particularly in Africa. United Nations, New York.

UNCCD Secretariat. (2013). A Stronger UNCCD for a Land-Degradation Neutral World, 20. Retrieved from http://sustainabledevelopment.un.org/content/documents/1803tstissuesdldd.pdf

UNDP (1992 Report) United Nations Development Programme. www.undp.org

United Nations Convention to Combat Desertification. (2015). Costs and benefits of policies and practices addressing land degradation and drought in the drylands: White Paper II.

Vermang, J., Gabriels, D., Cornelis, W., & De Boever, M. (Eds.). (2011). Land degradation processes and assessment: wind erosion, interrill erosion, gully erosion, land cover features. Ghent, Belgium: Ghent University. UNESCO Chair of Eremology; Research Foundation Flanders (FWO).

Venkateswarlu, B. & Prasad, J. V. N. S. (2011). Issues and Strategies for Managing Degraded Lands in Rainfed Ecosystem in India. In Dipak Sarkar, A. B. Azad, S. K. Singh & N. Akter (Eds.), Strategy for Arresting Land Degradation in South Asian Countries, (pp. 191–207). Dhaka, Bangladesh, SAARC Agriculture Centre.

Zaz, S. N. & Romshoo, S. A. (2012). Assessing the Geoindicators of Land Degradation in the Kashmir Himalayan Region. Natural Hazards, 64, 1219–1245. https://doi.org/10.1007/s11069-012-0293-3

Zdruli P., Pagliai, M., Kapur, S., and Cano, A. F. (2010a). What We Know About the Saga of Land Degradation and How to Deal with It? Chapter in Land Degradation and Desertification: Assessment, Mitigation and Remediation. By Pietsch, D., & Morris, M. https://doi.org/10.1007/978-90-481-8657-0

Zdruli, P., Pagliai, M., Kapur, S. & Cano, A. F. (2010b). What We Know About the Saga of Land Degradation and How to Deal with It?. In P. Zdruli, M. Pagliai, S. Kapur, & A. F. Cano (Eds.), Land Degradation and Desertification: Assessment, Mitigation and Remediation, (pp. 3–14). London, NY, Springer Dordrecht Heidelberg.

Chapter 2
Land Degradation: Indian Scenario

Abstract Total land degradation areas in India have been recorded over the areas of 4,67,021 km² in 2008–2009, while it has increased in 2006 with the srea of 4,72,262 km² under degraded land. 14.75% of total geographical areas are under different types of land degradation in 2008–2009. Similarly, the area under desertification in India has been increased by 1.16 million ha from 2003–2005 to 2011–2013. This chapter has tried to look into spatial and temporal understanding of land degradation types to get a glimpse of gravity of problems in the Indian scenario as India is not only a major agricultural dominating nation; however, it has witnessed stagnation in agricultural productivity in different regions like Punjab. The chapter has tried to understand land degradation through using methods like principal component analysis (PCA).

Keywords Land Degradation Index (LDI) · Wasteland · Gully erosion · Soil salinity · Principal component analysis (PCA) · Factor analysis

2.1 Introduction

Productive land and soil are critical natural capital assets essential for agricultural productivity, conserving biodiversity, and the provision of ecosystem services, such as carbon sequestration, water purification and storage, biofuels, climate protection and regulation, and natural heritage (UNCCD 2013). The declining land quality is termed as land degradation (Shroder et al. 2016). Land degradation is often the result of land mismanagement, including deforestation, overgrazing, monoculture, salinization, misuse of fertilisers and/or chemicals, poor farming practices, and soil erosion (Schauer 2014). Many land degradation types are found with the varying causes. It has direct implications on the ecosystem and the agricultural productivity. This phenomenon is affecting (directly or indirectly) billions of people around the world and may lead to the overexploitation of soil resources, loss of ecosystem productivity, shifts in vegetation composition, and/or loss of rural livelihoods (Shroder et al. 2016).

© The Author(s), under exclusive license to Springer Nature Switzerland AG 2021
R. Priya, *Land Degradation in India*, SpringerBriefs in Environmental Science,
https://doi.org/10.1007/978-3-030-68848-6_2

Land degradation is both a natural and human-induced process. It existed before the human race populated the earth and will continue to exist. However, humans have a two-sided effect on it: mitigate or accelerate (Zdruli et al. 2010). However, before mitigation or acceleration, understanding of land degradation is more necessary for all especially in developing countries (Gupta and Sharma 2010). Therefore, indicator-based approaches are often used to monitor land degradation and desertification from the global to the very local scale. Indicators are becoming increasingly important for communicating information to policymakers and the general public, as well as for assessing the environmental performance and the progress made by actions applied to mitigate land degradation and desertification (Kairis et al. 2013).

Therefore, this study will first try to understand the spatial distribution of land degradation types. Then the indexing of land degradation will be done, and its spatial extent will be seen according to agro-climatic regions. All these studies will be done with special focus on soil salinity.

2.2 Methodology and Database

2.2.1 Statistical Methods

Major espoused methodologies in this study are distance from mean and principal component analysis (PCA) for fabricating composite index for different indicators of land degradation: gullied or ravenous land (medium), gullied or ravenous land (deep), land with dense scrubs, land with open scrubs, land affected by salinity/alkalinity (medium), and land affected by salinity/alkalinity (strong). Sequentially, by both the method, Land Degradation Index (LDI) is computed, and comparative studies have been done between the two different methodologies: (i) the distance from mean and (ii) principal component analysis. Furthermore, linear regression model (OLS) (Pani and Carling 2013) has been adopted as a means of achieving the understanding about interrelationships between land degradation and unforested areas. Moreover, mean, standard deviation, and cross-tabulated data have been performed to elaborate the analysis in a way that is more coherent.

Factor analysis comes in performance when all variables are correlated to some extent. This analyzes observations of variables in three steps- computation of correlation matrix, extraction of initial factors and run finally rotation of the extracted factor to final solution as instructed in work of Ho (2006). First, principal component analysis is carried out on each dataset, which is then normalized through dividing all dataset by the square root of the first eigenvalue obtained from each PCA. Then linear function has been accounted, and Land Degradation Index is computed:

$$LDI = a1X1 + a2X2 + a3X3 + a4X4 \cdots + a6X6.$$

where a is the coefficient vector and X is the selected variables.

2.2.2   Database

District-wise status and spatial distribution of degraded and wasteland of India are made available by a collaboration of the National Remote Sensing Centre (NRSC) and Department of Land Resources (Ministry of Rural Development, GoI). The remote sensing technology has been used to provide data by earlier metioned instituions and it has provided data regarding land use and physical conditions of the surface features of agricultural and non-agricultural areas (presently non-arable) on a 1:50,000 scale using satellite images of Resourcesat-1 LISS-III for 2008–2009 to the major seasons like Kharif, Rabi, and Zaid with adequate field checks.

Wasteland data has been used of 2003–2005, 2008–2009, and 2011–2013 for the analysis. It has been taken from various reports of the National Remote Sensing Centre (NRSC) and Department of Land Resources (Ministry of Rural Development, GoI) (2011, 2016).

Forest Survey of India (Ministry of Environment & Forest, Government of India) provides data on forestation. Forest cover data has been used for 2009 taken from the report of FSI 2011 and 2015. FSI has used remote sensing data of IRS-1C/1D LISS III (Report FSI, 2001), IRS-P6-LISS III (Report FSI, 2009), IRS-P6-LISS III, and IRS-P6 AWiFS (Report FSI, 2011) with a 23.5 m resolution except IRS-P6 AWiFS (56 m resolution) on the scale of 1:50000 for extracting the data related to forest cover. The report of 2017 of FSI has been used for the explanation of forest cover of district understanding for the case study.

2.3   Land Degradation in India: Temporal and Spatial Extent

2.3.1   Extent of Land Degradation[1]

Total land degradation areas in India have been recorded over the areas of 4,67,021 km^2 in 2008–2009 (NRSC 2011), while it recorded an increase from 2005 to 2006 (4,72,262 km^2 degraded land in this year). Approximately 14.75% of total geographical areas are under different types of the land degradation in 2008–2009. The categorization of land degradation types has been done differently in the different year by the National Remote Sensing Agency, Government of India. Therefore, this study has tried to explain land degradation types in different years separately. The year 2008–2009 has recorded a decline in the degraded land by 5240.78 km^2 from 2005 to 2006 (NRSC 2011). Similarly, the area under desertification in India has been increased by 1.16 million ha from 2003–2005 to 2011–2013 (NRSC 2016). In 2005–2006 and 2008–2009, the share of "land with scrub" category of land degradation to total degraded area (TDA) is 39.17% and 38.54%, respectively, which is

---

[1] Priya (2018), Land Degradation and Indian Agriculture: A Regional Analysis. Unpublished thesis submitted to Jawaharlal Nehru University, New Delhi.

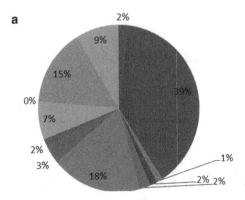

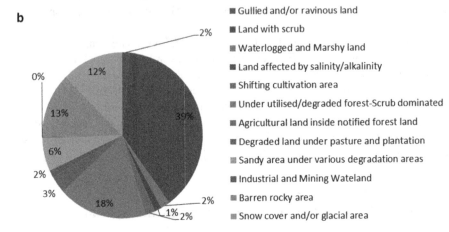

Fig. 2.1 (**a**) Share of land degradation types into total degraded land in 2005–2006. (**b**) Share of land degradation types into total degraded land in 2008–2009. (Source: NRSC Wasteland Atlas 2011)

highest among all categories (Fig. 2.1a, b). This is followed by underutilized/ degraded forest scrub dominated (17.92%), barren rocky area (12.74%), snow covered and/or glacial area (12.46%), sandy area under various degradation areas (6.48%), agricultural land inside notified forest land (3.36%), shifting cultivation (1.93%), waterlogged and marshy land (1.86%), gullied and/or ravinous land (1.59%), degraded land under pasture and plantation (1.52%), land affected by salinity/alkalinity (1.46%), and industrial and mining wasteland (0.14%) (Fig. 2.1b).

Similarly, NRSC (2016) has given different types of degradation explanation. Water erosion (~37%) accounts for the highest degradation into TDA in 2011–2013, and it has recorded a decrease from 2003 to 2005 when it was ~38% of TDA (Fig. 2.2). Vegetation degradation (~30% of TDA) and wind erosion (~19% of TDA) have not changed over the period of 2003–2005 to 2011–2013. Similar is the case with salinity (~4% of TDA), waterlogging (~1% of TDA), mass movement

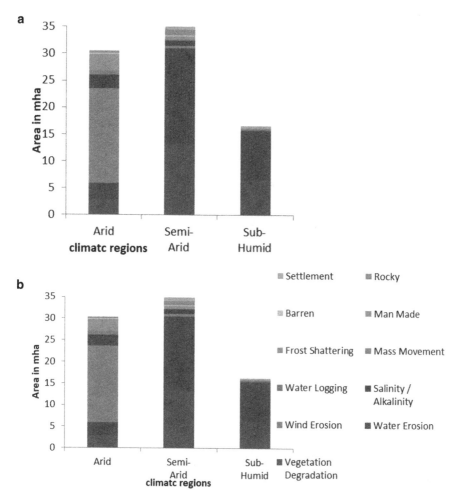

Fig. 2.2 (**a**) Land degradation types into various climatic region in 2011–2013. (**b**) Land degradation types into various climatic region in 2003–2005. (Source: NRSC Wasteland Atlas 2016)

(~1% of TDA), barren/rocky (~2% of TDA), and settlement (~2% of TDA). On the other hand, frost shattering has witnessed an increase of ~1% of TDA from 2003–2005 to 2011–2013 (Fig. 2.3a, b).

The climatic region-wise analysis reveals that the semi-arid region has the highest share of degradation into TDA amounting 34.85% and 35.4% in 2003–2005 and 2011–2013, respectively (Table 2.1). This is followed by the arid region and sub-humid region (Table 2.1). The arid region has experienced wind erosion heavily accounting 17.63 million ha (mha) of land in 2011–2013. In the same region, the degradation types like water erosion (3.03 mha), frost shattering (2.94 mha), vegetation degradation (2.86 mha), and salinity and alkalinity (2.52 mha) are very much prominent, among others (in 2011–2013) (Table 2.1 and Fig. 2.2a).

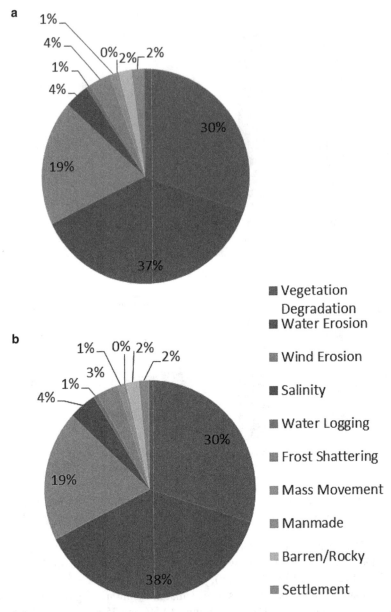

Fig. 2.3 (**a**) Land degradation types in 2011–2013. (**b**) Land degradation types in 2003–2005. (Source: NRSC Wasteland Atlas 2016)

Table 2.1 Area under different process of land degradation

Process of degradation	Area under desertification (mha)							
	2011–13				2003–05			
	Arid	Semi-arid	Sub-humid	Total	Arid	Semi-arid	Sub-humid	Total
Vegetation Degradation	2.86	13.48	6.65	22.99	2.81	13.39	6.34	22.55
Water erosion	3.03	17.51	8.97	29.51	3.12	17.07	8.91	29.11
Wind erosion	17.63	0.56	0	18.19	17.72	0.57	0	18.3
Salinity/alkalinity	2.52	0.86	0.09	3.48	2.52	1.07	0.21	3.8
Waterlogging	0.02	0.08	0.31	0.42	0.02	0.08	0.25	0.36
Mass movement	0.84	0.11	0	0.96	0.76	0.11	0	0.87
Frost shattering	2.94	0.46	0.01	3.41	2.74	0.43	0.01	3.18
Man-made	0.04	0.14	0.16	0.35	0.04	0.14	0.14	0.32
Barren	0.25	0.28	0.05	0.58	0.25	0.28	0.05	0.58
Rocky	0.3	0.97	0.02	1.29	0.29	0.97	0.02	1.28
Settlement	0.11	0.93	0.44	1.47	0.07	0.75	0.33	1.15
Grand total	**30.54**	**35.4**	**16.7**	**82.64**	**30.35**	**34.85**	**16.28**	**81.48**

Source: NRSC Wasteland Atlas (2016)

However, the semi-arid region has gone through water erosion significantly reporting 17.51 mha (2011–13) area has suffered from it. This is followed by the vegetation degradation (13.48 mha), rocky areas (0.97 mha), settlement (0.93 mha), salinity/alkalinity (0.86 mha), wind erosion (0.56 mha), etc. (Table 2.1 and Fig. 2.2a).

Figure 2.4a has observed that the arid region has recorded an increase into vegetation degradation (increased by 0.05 mha), mass movement (increased by 0.08 mha), frost shattering (increased by 0.20 mha), rocky area (increased by 0.01 mha), and settlement (increased by 0.04 mha). However, water erosion (decreased by 0.09 mha) and wind erosion (decreased by 0.09 mha) have witnessed a decrease in the arid region. In addition, the degradation types like salinity/alkalinity, waterlogging, man-made degradation, and barren land have recorded no change/negligible change (Fig. 2.4a). Overall, the arid region has registered a growth in total land degradation by 0.19 mha. Interestingly, the semi-arid region (increase by 0.55 mha land degradation area) is witnessing more increase in overall land degradation compared to the arid (increased by 0.19 mha) and sub-humid regions (increased by 0.42 mha) from 2003–2005 to 2011–2013 (Table 2.1).

The semi-arid region has seen an increase into vegetation degradation (increased by 0.09 mha), water erosion (increased by 0.44 mha), frost shattering (increased by 0.03 mha), and settlement (increased by 0.18 mha). Water erosion (decreased by 0.01 mha) and salinity/alkalinity (decreased by 0.21 mha) have witnessed a decline into the semi-arid region. Indeed, waterlogging, mass movement, man-made degradation, barren land, and rocky land have not indicated any change from 2003 to 2005 (Fig. 2.4b). Figure 2.4c has revealed about changes in land degradation areas into the sub-humid region. It is brought out that vegetation degradation, water erosion, waterlogging, man-made degradation, and settlement have recorded an

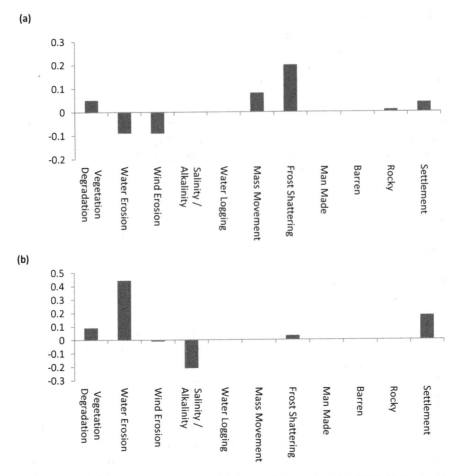

Fig. 2.4 (**a**) Arid region: land degradation change from 2003–2005 to 2011–2013. (**b**) Semi-arid region: land degradation change from 2003–2005 to 2011–2013. (**c**) Sub-humid region: land degradation change from 2003–2005 to 2011–2013. (**d**) Aggregate: land degradation change from 2003–2005 to 2011–2013. (Source: Prepared by the author by using data from the Wasteland Atlas of India, NRSC 2016)

increase by 0.31 mha, 0.06 mha, 0.06 mha, 0.02 mha, and 0.11 mha, respectively. Conversely, only salinity/alkalinity has witnessed a decrease by 0.12 mha (Table 2.1).

Overall in these regions, there is a change of 1.16 mha in land degradation from 2003–2005 to 2011–2013. The total increase in different types of land degradation into these areas is vegetation degradation by 0.44 mha, water erosion by 0.40 mha, waterlogging by 0.06 mha, mass movement by 0.09 mha, frost shattering by 0.23 mha, man-made degradation by 0.03 mha, rocky land by 0.01 mha, and settlement by 0.32 mha. Wind erosion (by 0.11 mha) and salinity/alkalinity (by 0.32 mha) areas have declined totally. No change was recorded in the barren land into total climatic regions (Table 2.1 and Fig. 2.4d).

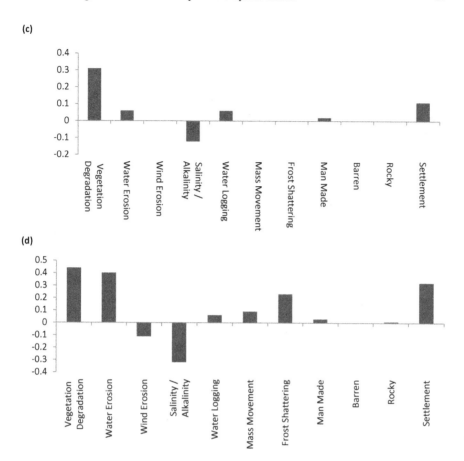

Fig. 2.4 (continued)

Table 2.2 has explained the status of land degradation into the various states of India. The average land degradation in the whole India is 14.75% of the total geographical areas (TGA). Ten states have recorded more than the average of India (proportionate to TGA). These states are Jammu and Kashmir (74.4% of TGA), Sikkim (46.13% of TGA), Himachal Pradesh (40.14% of TGA), Nagaland (31.77% of TGA), Manipur (25.30% of TGA), Rajasthan (24.82% of TGA), Uttarakhand (24.04% of TGA), Mizoram (23.52% of TGA), Meghalaya (18.40% of TGA), and Arunachal Pradesh (17.79% of TGA). Besides, other states have recorded less than the average of India; notably, they are still having significant areas under land degradations.

The changes from 2005–2006 to 2008–2009 are very noticeable. Eight states have registered an increase in areas under different land degradation types, and these states are Arunachal Pradesh (increase of 9151.41 km², 10.93% of TGA), Bihar (2759.92 km², 2.93% of TGA), Jammu and Kashmir (1681.39 km², 1.66% of

Table 2.2 Status of land degradation in states from 2005–2006 to 2008–2009

State name	Total wasteland (hectare)					% to TGA		
	2005–2006	2008–2009	Change	Total reduction	Total increase	2005–2006	2008–2009	% change over 2005–2006
Andhra Pradesh	38788.22	37296.62	−1491.6	1682.1	190.46	14.1	13.56	−0.54
Arunachal Pradesh	5743.83	14895.24	9151.41	108.48	9259.89	6.86	17.79	10.93
Assam	8778.02	8453.86	−324.15	862.56	538.04	11.19	10.78	−0.41
Bihar	6841.09	9601.01	2759.92	1895.09	4654.41	7.26	10.2	2.93
Chhattisgarh	11817.82	11482.18	−335.64	379.06	43.15	8.74	8.49	−0.25
Delhi	83.34	90.21	6.87	3.62	10.27	5.62	6.08	0.46
Goa	496.27	489.08	−7.18	11.48	3.99	13.41	13.21	−0.19
Gujarat	21350.38	20108.06	−1242.32	2858.99	1616.67	10.89	10.26	−0.63
Haryana	2347.05	2145.98	−201.07	232.2	31.92	5.31	4.85	−0.45
Himachal Pradesh	22470.05	22347.88	−122.17	197.25	75.57	40.36	40.14	−0.22
Jammu and Kashmir	73754.38	75435.77	1681.39	1191.48	2872.78	72.75	74.4	1.66
Jharkhand	11670.14	11017.38	−652.76	1183.5	531.16	14.64	13.82	−0.82
Karnataka	14438.12	13030.62	−1407.5	1477.98	70.82	7.53	6.79	−0.73
Kerala	2458.69	2445.62	−13.07	247.55	234.44	6.33	6.29	−0.03
Madhya Pradesh	40042.98	40113.27	70.29	258.95	329.25	12.99	13.01	0.02
Maharashtra	38262.81	37830.82	−431.99	469.93	38.22	12.44	12.3	−0.14
Manipur	7027.47	5648.53	−1378.94	2391.1	1012.14	31.48	25.3	−6.18
Meghalaya	3865.76	4127.43	261.67	93.86	355.13	17.24	18.4	1.17
Mizoram	6021.14	4958.64	−1062.5	2669.27	1606.71	28.56	23.52	−5.04
Nagaland	4815.18	5266.72	451.55	721.37	1172.6	29.04	31.77	2.72
Orissa	16648.27	16425.76	−222.51	271.75	48.69	10.69	10.55	−0.14
Punjab	1019.5	936.83	−82.67	112.7	30.56	2.02	1.86	−0.16
Rajasthan	93689.47	84929.1	−8760.37	10264.6	1503.37	27.38	24.82	−2.56
Sikkim	3280.88	3273.15	−7.73	11.83	4.29	46.24	46.13	−0.11
Tamil Nadu	9125.56	8721.79	−403.77	426.78	22.74	7.02	6.71	−0.31
Tripura	1315.17	964.64	−350.53	486.15	135.07	12.54	9.2	−3.34
Uttarakhand	12790.06	12859.53	69.47	440.35	509.86	23.91	24.04	0.13
Uttar Pradesh	10988.59	9881.24	−1107.35	1269.71	163.08	4.56	4.1	−0.46
West Bengal	1994.41	1929.2	−65.21	92.98	28.46	2.25	2.17	−0.07
Union Territory	337.3	315	−22.3	27.33	4.68	3.55	3.32	−0.23
Total	472261.9	467021.2	−5240.78	32,340	27098.43	14.91	14.75	−0.17

Source: NRSC Wasteland Atlas (2011)

TGA), Nagaland (451.55 km², 2.72% of TGA), Meghalaya (261.67 km², 1.17% of TGA), Madhya Pradesh (70.29 km², 0.02% of TGA), Uttarakhand (69.47 km², 0.13% of TGA), and Delhi (6.87 km², 0.46% of TGA) (Table 3.2). Overall, India has witnessed a decline in the areas of degradation by 5240.78 km², which accounts for 0.17% of TGA. This is indicative that the majority of states have recorded the decline, in which the major states are Rajasthan (decline by 8760.37 km², 2.56% of TGA), Andhra Pradesh (1491.6 km², 0.54% of TGA), Karnataka (1407.5 km², 0.73% of TGA), Manipur (1378.94 km², 6.18% of TGA), Gujarat (1242.32 km², 0.63% of TGA), Uttar Pradesh (1107.35 km², 0.46% of TGA), etc. (Table 2.2).

As many as 260 districts are affected by the gully/ravines (medium) in India covering 6145.96 km² areas in 2008–2009. The districts like Morena (498.25 km²) and Bhind (422.57 km²) of Madhya Pradesh are having highest areas under gully/ravines, which are reportedly about 10% and 9.5% of TGA of the district, respectively. Districts like Morena (9.98% of TGA), Bhind (9.48% of TGA), Bilaspur (5.46% of TGA), Rajouri (3.97% of TGA), Jalaun (3.37% of TGA), Perambalur (3.37% of TGA), and Una (3.04% of TGA) are affected by intense gully/ravine (medium) (Fig. 2.5a). The districts such as Rajouri (2.25% of TGA), Bundi (3.44% of TGA), Firozabad (3.01% of TGA), Kathua (2.6% of TGA), Dhaulpur (1.02% of TGA), Etawah (2.21% of TGA), Agra (1.46% of TGA), and Auriya and Leh (both ~1% of TGA) are highly affected by gully/ravines (deep). A total of 51 districts are affected by the gully/ravines (deep) accounting 1266.06 km² (Fig. 2.5b). Overall, 273 districts are being affected by one of both types and both types of gully/ravines. Morena, Bhind, Rajouri, Bilaspur, Bhind, Firozabad, Kathua, Jalaun, and Dhaulpur are highly affected by very high gully/ravine type of land degradation (Fig. 2.5c). Interestingly, only one very highly gully/ravine-affected district has witnessed both the medium and deep ravines, which is Rajouri district of Jammu and Kashmir (Fig. 2.5).

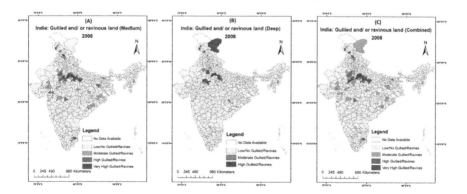

Fig. 2.5 (**a**) Spatial pattern of gully and ravines land (medium) in India. (**b**) Spatial pattern of gully and ravine land (deep) in India. (**c**) Spatial pattern of gully and ravine land (combined) in India. (Source: Prepared by the author by using data provided by NRSC through Wasteland Atlas 2011)

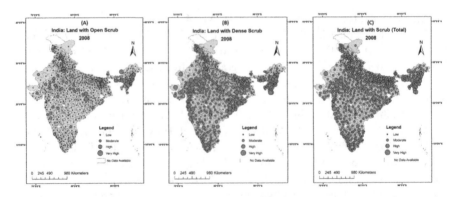

Fig. 2.6 (**a**) Spatial pattern of land with open scrub in India. (**b**) Spatial pattern of land with dense scrub in India. (**c**) Spatial pattern of land with scrub (total) in India. (Source: Prepared by the author by using data provided by NRSC through Wasteland Atlas 2011)

Surface run-off mismanagement, especially deforestation, overgrazing, and unsuitable farming practices are attributed as causes of gully erosion by various literatures. The pace of developing gully usually intensifies due to concentration of rainfall during the monsoon. Specifically, the relative significance of climatic factors in ravine formation in India has been a source of controversy. However, most of the studies favour multiple combinations of socio-economic and biophysical factors rather than any single set of factors responsible for ravine formation (Sharma 1980; Pani et al. 2011; Pani and Carling 2013; Priya 2014a, b; Priya and Pani 2015).

More than 524 districts are being affected by the dense scrub covering the area of 86979.91 km^2 in 2008–2009 (Fig. 2.6a). Land with open scrub problem exists into almost every district (covering 93033.00 km^2) except 17 districts like Baghpat, Chennai, Auraiya, island territories, Dakshin Dinajpur, Nadia, North 24 Parganas, Hugli, Kolkata, Haora, South 24 Parganas, Thiruvarur, Mansa, Mumbai, Nalbari, and Maharajganj (Fig. 2.6b). Similarly, overall land affected with scrubs is spread across India except for the same 16 districts except Baghpat (Fig. 2.6b). The whole area under the scrub was 180012.91 km^2 in 2008–2009. Districts like Korba, Senapati, Rajsamand, Churachandpur, Hamirpur, Jaisalmer, Mon, Tawang, North Cachar Hills, etc., are very highly affected by scrubs (total) (Figs. 2.6c and 2.7).

According to the proportionate to TGA, the high waterlogged/marshy land (permanent) districts are Puri, Purba Champaran, and Katihar. Area-wise scenario is the same with the area of 133.37 km^2, 77.33 km^2, and 58.68 km^2, respectively. 198 districts are being affected by waterlogging/marshy land, which has covered 1757.07 in km^2 2008–2009 (NRSC 2011). On the other hand, areas under waterlogged and marshy land (seasonal) are 6946.31 km^2 spreading across 211 districts (NRSC 2011). Overall, the district of Munger (24% to TGA) is highly affected by waterlogging and marshy land, and districts like Madhepura, Nawada, Saharsa, Buxar, Bhagalpur, Gopalganj, Khagaria, and Nalanda are moderately affected by waterlogging and marshy land problem, each accounting more than 10% to TGA (Fig. 2.8c).

Fig. 2.7 (**a**) Google Earth image of Pachpadra Lake of Barmer District of Rajasthan (May 20, 2016). The red colour demarcates the area of the Pachpadra Lake, and the yellow colour shows the dumping of waste into the lake. (**b**) Field photo of Pachpadra Lake. Image is taken from inside the Lake (Priya (2018), Land Degradation and Indian Agriculture: A Regional Analysis. Unpublished thesis submitted to Jawaharlal Nehru University, New Delhi.)

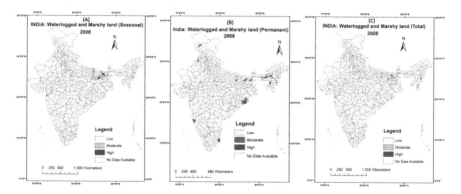

Fig. 2.8 (**a**) Spatial pattern of waterlogged and marshy land (seasonal) in India. (**b**) Spatial pattern of waterlogged and marshy land (permanent) in India. (**c**) Spatial pattern of waterlogged and marshy land (total) in India. (Source: Prepared by the author by using data provided by NRSC through Wasteland Atlas 2011)

India is estimated to have about 58.2 million ha of wetlands, many of which are distributed around the Indo-Gangetic plains, and it is generally regarded as "a water-surplus area". Waterlogging, closely associated with salinization and/alkalinization, continues to be a threat to sustained irrigated agriculture (Pandey et al. 2010).

The soil salinity/alkalinity problem is also being one of the prominent land degradation types in India. 168 districts are being affected by salinity/alkalinity (moderate/medium) spreading over 5414.53 km² area, accounting for 0.17% of TGA (NRSC 2011). Major districts like Kannauj, Raebareli, Auraiya, Mainpuri, Kanpur Rural, Unnao, Kanpur Urban, etc., are affected by the salinity/alkalinity (moderate/medium) (Fig. 2.9a). Similarly, 104 districts are affected by salinity/alkalinity

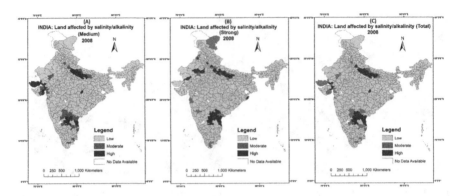

Fig. 2.9 (**a**) Spatial pattern of land affected by salinity/alkalinity (medium) in India. (**b**) Spatial pattern of land affected by salinity/alkalinity (Strong) in India. (**c**) Spatial pattern of land affected by salinity/alkalinity (total) in India. (Source: Prepared by the author by using data provided by NRSC through Wasteland Atlas 2011)

(strong) covering the areas of 1391.09 km². Major districts are so many, which are highly affected by salinity/alkalinity (strong), including Pratapgarh, Unnao, Sultanpur, Lucknow, Kannauj, Kurnool, Sonipat, etc. (Fig. 2.9b).

Soil salinity problem has been the resultant of multitude of factors, such as inefficient canal irrigation, shortfall in the surface and subsurface drainage, inadequate and inefficient water management, poor water supply system, extraction of poor quality of groundwater, poorly balanced water distribution, faulty irrigation and farming practices (Datta et al. 2000; Datta and Jong 2002; Ritzema et al. 2008), inappropriate cultural practices (Quereshi et al. 2008), and population growth and its pressure on land (Wicke et al. 2011). The traditional cause of salinity lies with the high groundwater level due to its ability to reach to the root zone, bringing salinity to the surface; conversely, discriminately pumping out the groundwater for irrigation has become a new threat to sodication and salinization of soil (Datta and Jong 2002).Consequently, interruption took place in the natural equilibrium of the input and output of the groundwater leading to seepage and percolation losses (Datta et al. 2000; Ritzema et al. 2008).

As a whole, the highly salinity/alkalinity-affected districts are Kannauj, Raebareli, Unnao, Auraiya, Mainpuri, Pratapgarh, Kanpur Rural and Urban, Sultanpur, Lucknow, Etah, Etawah, Fatehpur, Hardoi, Patan, Firozabad, Jaunpur, Rajnandgaon, Farrukhabad, Nellore, Kaushambi, Ahmadabad, Sant Ravidas Nagar, Kurnool, Anantapur, Prakasam, etc. (Fig. 2.9c).

Shifting cultivation has affected the quality of the soil, therefore impacting the productivity in negative ways (Lindskog and Tengberg 1994; Reddy et al. 2002; Yao et al. 2013). India's significant portion of the areas is under the two different categories of shifting cultivation. The total 79 districts are under the category of shifting cultivation – current jhum – spreading over 4814.68 km² (Fig. 2.10a) (NRSC 2011). In addition, 70 districts are under the abandoned jhum extending over 4210.46 km² (Fig. 2.10b) (NRSC 2011). In India, total 9025.14 km² areas are under shifting

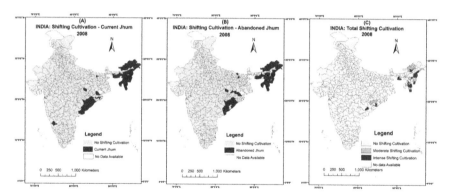

Fig. 2.10 (**a**) Spatial pattern of shifting cultivation – current jhum – in India. (**b**) Spatial pattern of shifting cultivation – abandoned jhum – in India. (**c**) Spatial pattern of total shifting cultivation in India. (Source: Prepared by the author by using data provided by NRSC through Wasteland Atlas 2011)

cultivation. The intensity of the shifting cultivation is high in the northeastern region and some districts of Odisha (Fig. 2.10c). Highly intense shifting cultivation areas are found in districts like Tirap (27.40% of TGA), Mon (22.13% of TGA), Tuensang (21.26% of TGA), Mokokchung (18.84% of TGA), Champhai (15.68% of TGA), Serchhip (12.81% of TGA), Zunheboto (12.53% of TGA), Gajapati (10.68% of TGA), etc. (Fig. 2.10c).

The underutilized forest is the major cause of concerns for the sustainable and optimum use of resources globally. India too is facing this problem through different intensities in different areas (Balooni and Singh 2003). Underutilized/degraded forests under agriculture are found prominently in more than 300 districts spreading over 15680.26 km² in 2008–2009 (NRSC 2011). Under it, districts like Korba, Sitamarhi, Kabeerdham, Golaghat, Kokrajhar, etc., are highly degraded (Fig. 2.11a). Besides, there is another category of degraded forest. This is degraded forest under scrub domination, which is found across more than 450 districts of India, accounting 83699.71 km² under the same degradation types (NRSC 2011). Korba, Srinagar, Sitamarhi, Kabeerdham, Karauli, Sheopur, Neemuch, Barwani, etc., are some districts which are accounting more than 15% of TGA under the underutilized/degraded forest (scrub domination) (Fig. 2.11b). Indeed, after combining both the category of the underutilized/degraded forest, Korba district has recorded as high as 72.71% area to TGA under it. It is followed by Sitamarhi (59% to TGA), Kabeerdham (40.55% to TGA), Srinagar (25.19% to TGA), Karauli (20% to TGA), and Sheopur and Kalan (19.67% to TGA), etc., which are under highly degraded forests (Fig. 2.11c).

Degraded pasture/grazing land has dominated in the arid and semi-arid regions as well as some parts of hilly areas. 6832.17 km² areas are under it spreading over 135 districts. Very highly degraded pasture/grazing land in India is found in districts like Erode, Idukki, Sonipat, Jhajjar, Rohtak, Bageshwar, Jind, Jodhpur, etc. (Fig. 2.12a). Degraded land under plantation crops is not so prominent in the

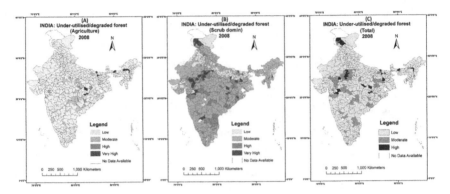

Fig. 2.11 (**a**) Spatial pattern of underutilized/degraded forest (agriculture) in India. (**b**) Spatial pattern of underutilized/degraded forest (scrub domination) in India. (**c**) Spatial pattern of total underutilized/degraded forest in India. (Source: Prepared by the author by using data provided by NRSC through Wasteland Atlas 2011)

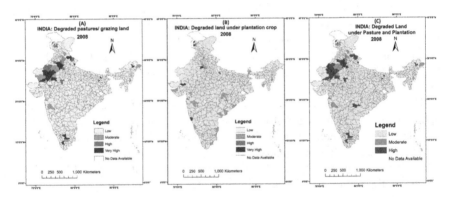

Fig. 2.12 (**a**) Spatial pattern of degraded pasture/grazing land in India. (**b**) Spatial pattern of degraded land under plantation crop in India. (**c**) Spatial pattern of degraded land under pasture and plantation in India. (Source: Prepared by the author by using data provided by NRSC through Wasteland Atlas 2011)

country. It accounts for only 278.53 km², and prominent districts under it are Kupwara (1.58% to TGA) and Mahendragarh (1.35% to TGA) (Fig. 2.12b). Moreover, total highly degraded land under pasture and plantation is found in districts like Erode, Karimnagar, Mahendragarh, Idukki, Jhajjar, Sonipat, Rohtak, etc. (Fig. 2.12c).

Catchment erosion has been found in the riverine zone of the river. Sands-riverine is one of the important types of land degradation (NRSC 2011). This has been dominated into the areas of river basins covering 2111.96 km². The Indus, Ganga, and Brahmaputra basins are the major regions of this category of degradation. Districts like Vaishali, Leh, Una, Korba, Dhemaji, etc., are highly affected by this type of degradation problem (Fig. 2.13a: marked by blue colour). Sands-coastal is more

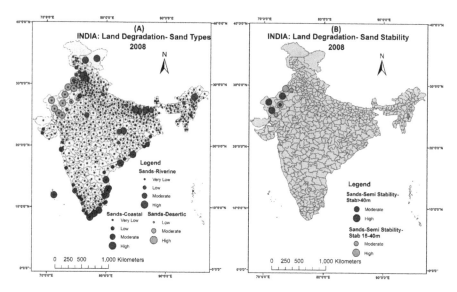

Fig. 2.13 (**a**) Spatial pattern of sands-riverine, sands-coastal, and sands-desertic in India. (**b**) Spatial pattern of sand stability in India. (Source: Prepared by the author by using data provided by NRSC through Wasteland Atlas 2011)

dominated into the east coast compared to the western coast of the Indian coastal region. Island territories are also affected by coastal sands. The districts which are highly affected under the sands-coastal are Lakshadweep, Ramanathapuram, Puducherry, Nellore, Kancheepuram, Thoothukudi, Kanyakumari, Srikakulam, East Godavari, Mumbai Suburban, etc. (Fig. 2.13a: marked by violet colour in the figure). The third type is sands-desertic, which is mostly dominated in Thar Desert and Punjab region. Districts like Jodhpur, Bikaner, Bhatinda, Mansa, Jalor, Muktsar, Ganganagar, Jaisalmer, Firozpur, Hanumangarh, etc., are highly affected under the desertic sand land degradation (Fig. 2.13a: marked by light green colour).

Shifting of sand is the major cause of desertification across the world and so is for India. Thus, sand stability has been a major indicator to assess the same. Sands-semi stable-stab > 40 m has been recorded only in five districts of Rajasthan-Jaisalmer, Barmer, Bikaner, Jodhpur, and Jalor (Fig. 2.13b: marked by violet colour). Moreover, sands-semi stable-stab 15–40 m is found highly concentrated in five districts of Bikaner, Jaisalmer, Ganganagar, Jodhpur, and Barmer. Other 15 districts majorly from northwestern India are having a moderate concentration of it (Fig. 2.13b: marked by light green colour).

Mining is emerging as a factor of degradation of land (Harden and Mathew 2000). The same problem has been encountered in different mining areas of eastern India region (Chhotanagpur, etc.), parts of northern India, some districts of southern India, and sparsely in western India. Highly affected districts of this categories are Faridabad, Cuddalore, North Goa, Rajsamand, Kannur, etc. (Fig. 2.14a). Industrial wastelands have been experienced in 61 districts covering areas of only 58 km^2.

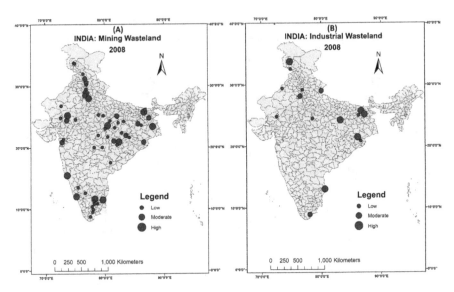

Fig. 2.14 (a) Spatial pattern of mining wasteland in India. (b) Spatial pattern of industrial wasteland in India. (Source: Prepared by the author by using data provided by NRSC through Wasteland Atlas 2011)

Highly affected districts are Chennai, Badgam, Bhagalpur, Sonbhadra, Kendujhar, and Thoothukudi (Fig. 2.14a).

NRSC (2011) has categorized barren/stony land and/or snow cover/glacial area under wasteland, which has extended to 59482.29 km^2 and 58183.44 areas, respectively. Both types of wasteland have dominated the western Himalayan region, and snow cover extends to the eastern Himalayan region too (Fig. 2.15a, b).

If the total degraded land is analysed, the ~60 districts of the country are such that more than 25% of areas to TGA are under different types of land degradation. Approximately 190 districts are such where 10.01–25% areas to TGA are being affected by land degradation. Moreover, ~140 districts are those whose 5.01–10% areas to TGA areas are being affected by land degradation. Approximately 182 districts are having degradation under <5% to TGA. Some of the major and highly degraded districts are Kargil, Leh, Lahaul and Spiti, North Sikkim, Kabeerdham, Jaisalmer, Sitamarhi, Kinnaur, Rajsamand, Udaipur, North Cachar Hills, Neemuch, Bikaner, Morena, etc. Moderately affected districts are Deoghar, Kohima, Pune, Pulwama, Jodhpur, Vishakhapatnam, Saharsa, Chittoor, Umaria, Kota, Banka, Ratnagiri, Anantpur, Kutch, Barmer, Nellore, Nashik, Vaishali, etc. The geomorphological distribution of land degradation is very uniquely set up. The highly degraded areas are concentrated in eastern Himalayan region, Purvanchal Hill Complex, Thar Desert, North-Central Plateau, near Nallamala Hills, and some parts of Chota Nagpur Plateau. Notably, the moderate land degradation areas are concentrated into the surrounding districts of the highly degraded districts with some exceptions (Fig. 2.16).

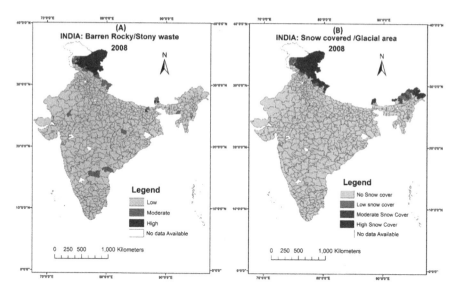

Fig. 2.15 (**a**) Spatial pattern of barren rocky/stony waste in India. (**b**) Spatial pattern of snow covered/glacial area in India. (Source: Prepared by the author by using data provided by NRSC through Wasteland Atlas 2011)

2.4 Land Degradation Index: Regional Approach

2.4.1 Land Degradation Index[2]

As the above is concerned on individual indicators, now it is similarly important to study aggregately. For observing composite index, distance from mean and principal component analysis (PCA) are calculated.

2.4.1.1 Result of Distance from Mean

In 2000, mean of land degradation is 517.00 km². Eleven districts are recorded under more degraded land than average. Its degradation is extremely high (more than 2000.00 km²). Kuchchh has accounted highest degradation with distance from mean of 12912.59 followed by Jaisalmer, Koraput, Malkangiri, Nabarangpur, Rayagada, Rajsamand, Udaipur, Nellore, Pune, and Jamnagar. Thirty-two districts have evidenced moderately high degradation. It is illustrated that most of these districts are round to the highly degraded districts. Figs. 2.16 and 2.17 are able to derive one interesting observation that high degradation is dominant in the arid and semi-arid regions. Milton et al. (1994) have observed that arid and semi-arid lands

---

[2]Priya (2014b), Unpublished dissertation is major source of this section, which is submitted to Jawaharlal Nehru University, New Delhi, Titled "Extent and Pattern and Implication of Land Degradation In India: Analysis of District-Wise Variation".

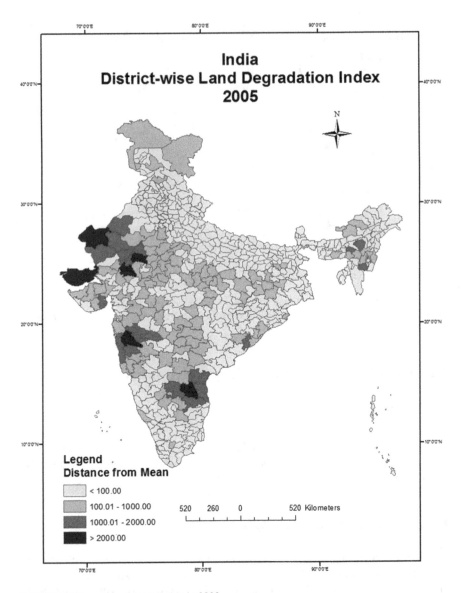

Fig. 2.16 Pattern of land degradation in 2000

have suffered loss of productivity and diversity during the twentieth century due to overuse of the rangeland. These areas come under low rainfall, and the variability in plants and animal production is 1.5 times greater than the variability in rainfall. Drought is a common phenomenon for arid and semi-arid areas. Thus, changes in practices and utilization of dry land are a significant solution to mitigate the impact of land degradation.

During year 2000–2008, only Jaisalmer and Pune districts of the severely degraded category in 2000 have brought more high degradation during the further years.

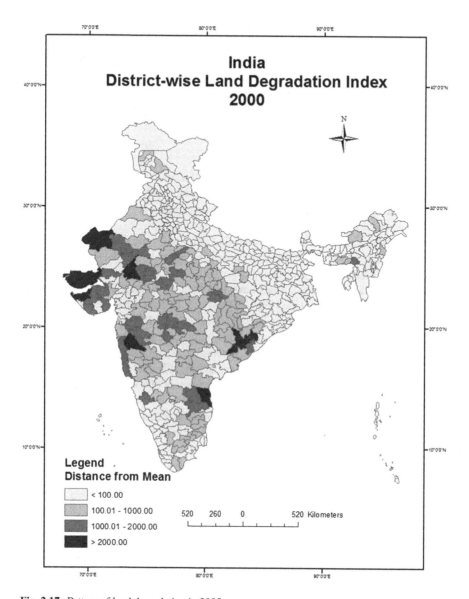

Fig. 2.17 Pattern of land degradation in 2005

Conversely, Malkangiri, Nabarangpur, and Rayagada (Orissa) have shown high improvement in degraded land. As a result, these come under low degraded category.

Koraput (Orissa), Nellore (Andhra Pradesh), and Jamnagar (Gujarat) also demonstrated decreasing trend, but it is relatively slow, and these come under moderately high-degraded land category. Bhilwara (Rajasthan) and Cuddapah (Andhra Pradesh) are initially under moderately high-degraded category, but 2005, these are recorded under severe degraded category. Moreover, in 2008 both districts come under highly degraded category diminutively. Chauhan (2003) has put forward

Table 2.3 Severely degraded districts by composite index by distance from mean

Year	Severely degraded districts (more than 2000 km^2 distance from mean)
2000	Kuchchh, Jamnagar (Gujarat), Jaisalmer, Rajsamand, Udaipur (Rajasthan), Koraput, Malkangiri, Nabarangpur, Rayagada (Orissa), Nellore (Andhra Pradesh), Pune (MH)
2005–2006	Jaisalmer, Udaipur, Bhilwara (Rajasthan), Kuchchh (Gujarat), Pune (MH), Cuddapah (Andhra Pradesh)
2008–2009	Jaisalmer, Udaipur (Rajasthan), Kuchchh (Gujarat), Pune (MH)

some solution of degraded land in desert areas that desertification must be controlled with adding "micro-climate" by planting shelter belts and wind break, which has really contributed to mitigate the hazards in desert areas (Priya 2014a, b (unpublished dissertation) (Table 2.3).

Above all the discussion has uncovered that although some districts appear with high increment in degraded land, overall land degradation decrements have occurred year by year. In such case, it is essential to focus on forest cover in our country. Initially it is narrated that India does have less than 33% forest cover of the total geographical areas. On one side, Figs. 2.16 and 2.17 have displayed that highly degraded districts are stepping down slowly. On another side, Fig. 2.15 has depicted that most of these districts have slightly or highly increasing trend from 2001 to 2011. For example, Jaisalmer, Bhilwara, Prakasam, Anantapur, Chittoor, Raigarh, Vijayanagaram, Barmer, Rajsamand, Ratnagiri, and Churachandpur are showing downward changes in degraded land with approximately 898 km^2, 50 km^2, 154 km^2, 82 km^2, 74 km^2, 37 km^2, 36 km^2, 18 km^2, 91 km^2, 345 km^2, 14 km^2, and 467 km^2, respectively. Out of these districts, decreasing trend in forest cover is recorded only in Cuddapah, Anantpur, Chittoor, and Ratnagiri. While forest cover of other districts shows an increasing tendency. This indicated that forest cover might be a useful solution to capture the problem related to land (Fig. 2.18).

2.4.1.2 Analysis of PCA Result

Siltation of dams, pollution of water courses by agricultural chemicals, damage to property by soil laden run-off, steep slopes, rippling terrain, faulty agricultural practices, and remotion of trees are many developmental activities that are responsible to compel the Earth at the risk of degradation (Maconachie 2007; Morgan 2006; Natarajan et al. 2010; Hong and Ju 2007; Clarke and Rendell 2007). Land Degradation Index has been calculated through distance from mean. Although this statistic is good to prepare composite index, principal component analysis (PCA) is better recommended for index. Mean and standard deviation of different indicators have showed that these indicators varied from each other (Table 2.4). Correlation matrix has indicated relation between indicators. Scrubs (dense) and scrubs (open), salinity (medium) and scrubs (dense), and salinity (strong and medium) are relatively highly correlated than other indicators. Other indicators are moderately correlated (Table 2.5).

Table 2.6 has evidenced eigenvalue, which demonstrates how much indicators are explained by this analysis. Six PCs are judged and are taken for study. First

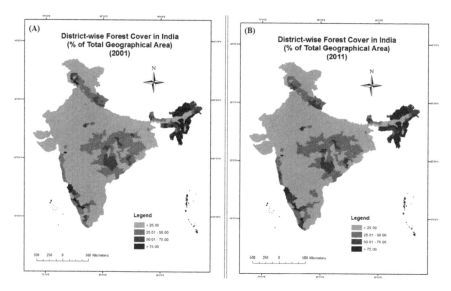

Fig. 2.18 Pattern of forest cover in 2001 (**a**) and 2011 (**b**)

Table 2.4 Descriptive statistics for the selected indicators for the study

Factors	Indicators	Mean	SD
X1	Gullied/ravine (deep)	10.72	35.800
X2	Gullied/ravine (medium)	2.20	18.366
X3	Scrubs (dense)	151.39	429.235
X4	Scrubs (open)	162.62	286.251
X5	Salinity (medium)	9.37	35.955
X6	Salinity (strong)	2.41	10.672

principal component has succeeded in representing approximately 36% of total variance; however, the first three principal components together have reported about 75% of total variance according to eigenvalues.

Relation between component and variables is explicated in component matrix (Table 2.7). The first three components have been selected for further study. According to analysis, the first PC has explained four variables, namely, scrubs (dense), scrubs (open), salinity (medium), and salinity (strong), very firmly. The second principal component has defined only gullied/ravine (medium) and salinity (strong) strongly. Like the second PC, the third PC has also showed only the first two components. Ultimately, the first component explains better to all variables in this case study.

Extents of variables explained have been accounted by getting the score of square of component loading. Thus, component loading has exhibited that loading of X3 on component one is 0.82. Therefore, the square of the loading is 67%. It indicates that component one is able to interpret 67% variance of X3 variable. Although component one has not explained the variables X1 and X2 with great extent, the attributes of variables X3, X4, X5, and X6 have been very well explained by 67%, 62%,

Table 2.5 Correlation matrix

Factors	X1	X2	X3	X4	X5	X6
X1	1.000					
X2	0.128	1.000				
X3	0.034	−0.003	1.000			
X4	0.075	0.027	0.770	1.000		
X5	0.046	0.023	0.321	0.251	1.000	
X6	0.052	0.167	0.194	0.160	0.540	1.000

Table 2.6 Eigenvalue of different indicators

Factor	Eigenvalue	% of variance	% of cumulative
X1	2.156	35.930	35.930
X2	1.247	20.784	56.714
X3	1.074	17.897	74.611
X4	0.867	14.443	89.054
X5	0.432	7.208	96.262
X6	0.224	3.738	100.000

Table 2.7 Component loadings

Component matrix	Component 1	Component 2	Component 3
X1	0.145	0.272	0.718
X2	0.131	0.533	0.515
X3	0.820	−0.436	0.118
X4	0.790	−0.455	0.214
X5	0.685	0.372	−0.375
X6	0.593	0.595	−0.304

46%, and 35%, respectively, in comparison to other components. After attentive examination, in spite of attending the basis of variable explained, there is also indigence to intention on the eigenvalues.

Considering Table 2.6, the eigenvalues after the first three components are diminishing very highly. As it is concerned already that first principal component is explaining approximately 40% of total variance in the present analysis, hence the equation for composite index will be

$$Y = 067 \ X_1 + .061 \ X_2 + .380 \ X_3 + .367 \ X_4 + .318 \ X_5 + .275 \ X_6$$

Coefficients of the different variables have been taken from score coefficients of the first principal component (Table 2.8).

According to the scores of the first principal component analysis, Land Degradation Index (LDI) has been prepared. Correspondingly, *four zones*[3] *of high degradation* have been determined (Fig. 2.19) as follows:

---

[3] Zone concept has been taken from regionalization concept of R. L. Singh.

Table 2.8 Component score coefficient matrix

Variables	PC1	PC2	PC3
X1	0.067	0.218	0.668
X2	0.061	0.427	0.480
X3	0.380	−0.350	0.110
X4	0.367	−0.364	0.20
X5	0.318	0.299	−0.350
X6	0.275	0.477	−0.283

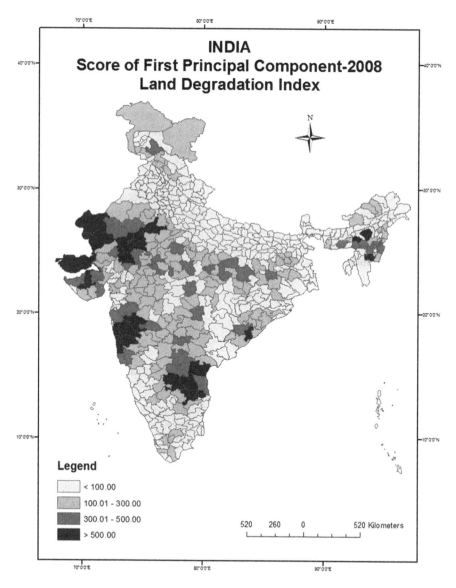

Fig. 2.19 Land Degradation Index according to the first PCA score

Thar Zone – Jaisalmer, Kuchchh, Udaipur, Bhilwara, Pali, Ajmer, Rajsamand, Barmer, Rajkot, etc.
North Deccan (Hot Semi-arid) Zone –Pune, Satara, Ratnagiri, Ahmadnagar, Jaipur etc.
South Deccan (Hot Semi-arid) Zone – Cuddapah, Anantapur, Prakasam, Chittoor, Vijayanagaram, etc.
Northeastern Zone– Karbi Anglong, Churachandpur, etc.[4,5]

2.5 Conclusion

The analysis has emphatically evidenced that Indian land is severally suffering from gullied/ravine, scrublands and salinity, and alkalinity. As it appeared that there are four zones of land degradation, these zones are not only highly vulnerable due to human beings (due to deforestation or unplanned irrigation and agriculture or mining activities) (Menon and Bawa 1998; Sahu and Dash 2011; Pani et al. 2011), but these zones come under agro-ecological zone. This has been evident that if negative changes in forestation arise out in large scale, it may result in environmental problem (Jolly 1993; Hong and Ju 2007), resources problem (Natarajan et al. 2010), and food insecurity (Tekle 1999; Reddy 2003; Lal 2009; Bhattacharya and Guleria 2012; Parsons 2014). Most significantly, this study brings out a very open relationship among land degradation with non-forested cover and population pressure, and data reveals that degradation of land takes place with the decreasing of forest cover land and increasing population density. To sum up, the need of strategy controlling deforestation, motivation to afforestation, and population control must be implemented to stop land degradation before it brings physical or financial ruin.

References

Anthopoulou, B., Panagopoulos, A., & Karyotis, Th. (2006). The Impact of Landscape in Northern Greece. Landslides, 3, 289–294. https://doi.org/10.1007/s10346-006-0056-x.
Balooni, K., & Singh, K. (2003). Financing of wasteland afforestation in India. Natural Resources Forum, 27(3), 235–246. https://doi.org/10.1111/1477-8947.00058.
Bhattacharya, T & Guleria, S. (2012). Costal Flood Management in Rural Planning Unit Through Land-Use Planning: West Bengal, India. Journal of Coastal Conservation, 16, 77–87. https://doi.org/10.1007/s11852-011-0176-x.
Chauhan, S. S. (2003). Desertification Control and Management of Land Degradation in the Thar Desert of India. The Environmentalist, 23(3), 219–227. https://doi.org/10.1023/B:ENVR.0000017366.67642.79
Clarke, M. L. & Rendell, H. M. (2007). Climate, Extreme Event and Land Degradation. In Mannava V. K. Sivakumar & Ndegwa Ndiang'ui (Eds.), Climate and Land Degradation, (pp.137–149). NY, Springer Berlin Heidelberg.

---

[4] Priya (2014b), the unpublished dissertation is major source of this section, which is submitted to Jawaharlal Nehru University, New Delhi, titled "Extent And Pattern and Implication of Land Degradation In India: Analysis of District-Wise Variation"
[5] Priya and Pani (2015)

Datta, K.K., Jong, C. D. and Singh, O.P. (2000). Reclaiming salt-affected land through drainage in Haryana, India: a financial analysis. Agricultural Water Management, 46, 55–71.

Datta, K. K. and Jong, C. de. (2002). Adverse effect of waterlogging and soil salinity on crop and land productivity in Northwest region of Haryana, India. Agricultural Water management, 57, 223–238.

Gupta, S. K., Ahmed, M., Hussain, M., Pandey, A. S., Singh, P., Saini, K. M…Das, S.N. (1998). Inventory of Degraded Lands of Palamau District, Bihar- A Remote Sensing Approach. Journal of the Indian Society of Remote Sensing, 26 (4), 161–168.

Gupta, S., & Sharma, R. K. (2010). Land Degradation – Its Extent and Determinants in Mountainous Regions of Himachal Pradesh §. Agricultural Economics Research, 23(June), 149–156.

Harden, C. P., & Mathews, L. (2000). Rainfall response of degraded soil following reforestation in the Copper Basin, Tennessee, USA. Environmental Management, 26(2), 163–174. https://doi.org/10.1007/s002670010079

Hong, M. & Ju, H. (2007). Status and Trends in Land degradation in Asia. In Mannava V. K. Sivakumar & Ndegwa Ndiang'ui (Eds.), Climate and Land Degradation, (pp.55–64). NY, Springer Berlin Heidelberg.

Ho, R. (2006). Handbook of Univariate and Multivariate Data Analysis and Interpretation with SPSS. https://doi.org/10.1201/9781420011111

Jolly, C. L. (1993). Population Change, Land Use and the Environment. Reproductive Health Matters, 1, 13–25. Retrieved from http://www.jstor.org/stable/3774852

Kairis, O., Kosmas, C., Karavitis, C., Ritsema, C., Salvati, L., Acikalin, S., Ziogas, A. (2013). Evaluation and Selection of Indicators for Land Degradation and Desertification Monitoring: Types of Degradation, Causes, and Implications for Management. Environmental Management, 54(5), 971–982. https://doi.org/10.1007/s00267-013-0110-0

Lal, R. (2009). Soil Degradation as a Reason for Inadequate Human Nutrition. Food Security, 1, 45–57. https://doi.org/10.1007/s12571-009-0009-z.

Lindskog, P. & Tengberg, A. (1994). Land Degradation, Natural Resources and Local Knowledge in the Sahel Zone of Burkina Faso. GeoJournal, 33 (4), 365–375. Retrieved from http://www.jstor.org/stable/41146235

Maconachie, R. (2007). Urban Growth and Land Degradation in Developing Cities: Changes and Challenges in Kano Nigeria. Hampshire, England: Ashgate Publication Limited.

Menon, S. & Bawa, K. S. (1998). Deforestation in the Tropics: Reconciling Disparities in Estimates for India. Amibio, 27 (7), 576–577. Retrieved from http://www.jstor.org/stable/4314794.

Morgan, R. P.C. (2006). Managing Sediment in the Landscape: Current Practices and Future Vision. In P. N. Owens (Eds.), Soil Erosion and Sediment Redistribution in River Catchments, (pp.287–293). Oxfordshire, UK, Biddles Ltd, King's Lynn.

Milton, S. J., Dean, W. R. J., Plessis, M. A. D. & Siegfried, W. R. (1994). Conceptual Model of Arid Rangeland Degradation. BioScience, 44 (2), 70–76. Retrieved from http://www.jstor.org/stable/1312204

Natarajan, A., Janakiraman, M., Manoharan, S., Kumar, K. S. A., Vadivelu, S. & Sarkar, D. (2010). Assessment of Land Degradation and Its Impacts on Land Resources of Sivagangai Block, Tamil Nadu, India. In P. Zdruli, M. Pagliai, S. Kapur, & A. F. Cano (Eds.), Land Degradation and Desertification: Assessment, Mitigation and Remediation, (pp. 235–252). London, NY, Springer Dordrecht Heidelberg.

NRSC & Department of Land Resources. (2011). District and Category-Wise Wastelands of India. Ministry of Rural Development, New Delhi, GOI & Indian Space Research Organization, Hyderabad. URL- http://www.dolr.nic.in/WastelandsAtlas2011/Wastelands_Atlas_2011.pdf

NRSC (2016). Wasteland Atlas of India. Government of India.

Parsons, A. J. (2014). Abandonment of Agricultural Land, Agricultural Policy and Land Degradation in Mediterranean Europe. In E. N. Mueller, J. Wainwright, A. J. Parsons & L. Turnbull (Eds.), Patterns of Land Degradation in Drylands: Understanding Self-Organised Ecogeomorphic Systems, (pp.357–366). London, NY, Springer Berlin Heidelberg.

Pani, P. & Carling, P. (2013). Land Degradation and Spatial Vulnerabilities: A Study of Inter-Village Differences in Chambal Valley, India. Asian Geographer, 30 (1), 65–79. Retrieved from https://doi.org/10.1080/10225706.2012.754775

Pani, P., Mishra, D. K. & Mohapatra, S. N. (2011). Land Degradation and Livelihoods in Semiarid India: A Study of Formers' Perception in Chambal Valley. Asian Profile, 39 (5), 505–519.

Pandey, A. C., Singh, S. K., & Nathawat, M. S. (2010). Waterlogging and flood hazards vulnerability and risk assessment in Indo Gangetic plain. Natural Hazards, 55(2), 273–289. https:// doi.org/10.1007/s11069-010-9525-6

Pimentel, D., Dazhong, W., Sanford, E., Lang, H., Emerson, D. & Karasik, M. (1986). Deforestation: Interdependency of Fuelwood and Agriculture. Oikos, 46 (3), 404–412. Retrieved from http:// www.jstor.org/stable/3565841

Priya (2018), Land Degradation and Indian Agriculture: A Regional Analysis. Unpublished thesis submitted to Jawaharlal Nehru University, New Delhi.

Priya, R. and Pani, P. (2015). Land Degradation and Deforestation in India: A District Level Analysis. Annals of National Association f Geographer, India, XXXV (1), June, 50–70. ISSN: 0970-972X

Priya, R. (2014a). Extent, Pattern and Implication of Land Degradation in India: Analysis of District-wise variation. Unpublished dissertation submitted at the Jawaharlal Nehru University, New Delhi

Priya (2014b), Unpublished Dissertation is major source of this section, which is submitted to Jawaharlal Nehru University, New Delhi, Titled "Extent And Pattern and Implication of Land Degradation In India: Analysis of District-Wise Variation".

Quereshi, A.S., McCornick, P. G., Qadir, M. & Aslam, Z. (2008). Managing Salinity and Waterlogging in the Indus Basin of Pakistan. Agricultural Water management, 95, 1–10. https:// doi.org/10.1016/j.agwat.2007.09.014

Reddy, G. P. O., Maji, A K., Srinivas, C. V, & Velayutham, M. (2002). Geomorphological Analysis for Inventory of Degraded Lands in a River Basin of Basaltic Terrain Using Remote Sensing and GIS. Journal of the Indian Society of Remote Sensing, 30(1,&2), 15–31. https://doi. org/10.1007/BF02989973

Reddy, V. R. (2003). Land Degradation in India: extent, costs and Determinant. Economic Political Weekly, 38 (44), 4700–4713. Retrieved from http://www.jstor.org/stable/4414225

Ritzema, H. P., Satyanarayana, T. V., Raman, S. and Boonstra, J. (2008). Subsurface drainage to combat waterlogging and salinity in irrigated lands in India: Lessons learned in farmers' fields. Agriculture Water Management, 95, 179–189. https://doi.org/10.1016/j.agwat.2007.09.012

Tekle, K. (1999). Land Degradation Problem and Their Implications for Food Shortage in South Wello, Ethiopua. Environmental Management, 23 (4), 419–427.

Sahu, H. B. & Dash, S. (2011). Land Degradation due to Mining in India and its Mitigation Measures. 2nd International Conference on Environmental Science and technology, IPCBEE, 6, 132–136. Singapore, IACSIT Press.

Schauer, M. (2014). THE E CONOMICS OF Economics of Land Degradation Initiative : Practitioner' s Guide for sustainable land management ,, The Economics of Land Degradation.

Sharma, H. S. 1980. Ravine Erosion in India. New Delhi: Concept.

Shroder, J. F., Sivanpillai, R., D'Odorico, P., & Ravi, S. (2016). Chapter 11 – Land Degradation and Environmental Change. In Biological and Environmental Hazards, Risks, and Disasters (pp. 219–227). https://doi.org/10.1016/B978-0-12-394847-2.00014-0

UNCCD Secretariat. (2013). A Stronger UNCCD for a Land-Degradation Neutral World, 20. Retrieved from http://sustainabledevelopment.un.org/content/documents/ 1803tstissuesdldd.pdf

Wicke, B., Smeets, E., Dornburg, V., Vashev, B., Gaiser, T., Turkenburge, W. and Faaij, A. (2011). The global technical and economic potential of bioenergy from salt-affected soils. Energy Environment Science, 4 (2669). https://doi.org/10.1039/c1ee01029h

Yao, R., Yang, J., Gao, P., Zhang, J., & Jin, W. (2013). Determining minimum data set for soil quality assessment of typical salt-affected farmland in the coastal reclamation area. Soil and Tillage Research, 128, 137–148. https://doi.org/10.1016/j.still.2012.11.007

Zdruli, P., Pagliai, M., Kapur, S. & Cano, A. F. (2010). What We Know About the Saga of Land Degradation and How to Deal with It?. In P. Zdruli, M. Pagliai, S. Kapur, & A. F. Cano (Eds.), Land Degradation and Desertification: Assessment, Mitigation and Remediation, (pp. 3–14). London, NY, Springer Dordrecht Heidelberg.

Chapter 3
Land Degradation in India: Relationship with Deforestation and Population

Abstract The analysis has emphatically evidenced that Indian land is severally suffering from gullied/ravine, scrublands and salinity, and alkalinity. As it appeared that there are four zones of land degradation, these zones are highly vulnerable due to human factors (due to deforestation or unplanned irrigation and agriculture or mining activities) with the natural factors. This chapter has tried to understand the human-land degradation relation through regression analysis. Results uncover that degradation of land takes place with the decreasing forest cover land and increasing population density.

Keywords Land degradation · Wasteland · Deforestation · Population · Regression analysis · Non-forest land

3.1 Introduction

Generally, the depletion of both land and forest depends on the types of practices done by human beings in those particular areas. Deforestation occurred in many conditions, i.e., lower rainfall, higher latitude, older island, distant from zone of aerial tephra, low island, small island, and isolated island. These changes are responsible for land degradation. As population size increases, the need for food also becomes high. As a result, there is negative relationship between land degradation and population. Similarly, rising population causes the declining quality of land (Priya 2014; Priya and Pani 2015; Lamb 2011). India, where the population is mostly depended on agriculture, has the problem of land degradation that leads to food insecurity. Land degradation might be extremely problematic, as one cannot imagine (Tekle 1999).

Exploitation of natural resources also results in environmental and land degradation. Excessive economic activities are inadequate to arrest or limit the availability of natural resources due to misallocation of resources and excessive use of resources. Soil, forest, and air are natural resources, which are degraded due to biotic pressure on forests, waterlogging, salinization, deforestation, and improper management of irrigation and soil erosion caused by the lack of soil and water conservation activities. Both economic growth and population growth are responsible for land and forest degradation. This paper also inferred that both poverty and environmental degradation are closely associated through their activities. Environmental degradation is linked to the production or consumption activity of poor people. Resource degradation in India is a serious problem, such as drinking water, sanitation, indoor pollution, water pollution, air pollution, resources exploitation, etc. Open access of resources is another region of land degradation. The paper of Premchander et al. (2003) has worked on a semi-arid region of Kerala and focused differently on water depletion as a cause of degradation of other natural resources. Water is a resource needed for drinking, agriculture, and livelihood as well as livestock. Water is a power for livelihood, but if there is more availability of water in agricultural areas, then it may convert in saline. Degradation of agricultural land in both qualitative and quantitative manner affects the health and productivity of both animals and people. Forest depletion is also declining year to year, which results in degradation of forest and land. This paper deeply is concerned with the agriculture and livelihood and further informs that both small- and large-scale farmers invest their own resources in digging well and borings, which helps them in irrigation. It is also revealed that groundwater depends on the porosity and permeability of the ground. This infiltration of water has been influenced by the depletion of rocks in hilly areas. Depletion and degradation of natural resources affects the livelihood of human beings. In long-term solution, this paper suggested some proposal like people must understand the complex interaction between the use of resources and livelihood. Highland and lowland management are required, and cutting rock and digging the bore wells are strictly stopped for increasing level of groundwater. Long-term sustainability is closely linked with the management of all natural resources.

3.2 Methodology and Database

To perceive impaction of non-forested land and population density on land degradation, as very concerned issues in literatures, the following methods and data sets are used.

3.2.1 Statistical Methods

Further, the present study has selected 104 low, medium, and highly land degraded districts of different states, and in order to investigate the impact of non-forested land and population density on the situation of land degradation, linear regression model (Rao 2009) has been done:

$$LDI = \beta 0 + \beta 1 \ X1 + \beta 2 \ X2 + u$$

Here, LDI is Land Degraded Index, X1 is the non-forested areas (% of total geographical areas), and X2 is the population density of selected 104 districts for the study.

3.2.2 Source of Data

The Land Degradation Index has been calculated by using SPSS statistical software. Forest cover data have been provided by Forest Survey of India (Ministry of Environment and Forest), Govt. of India. Non-forested cover areas have been extracted by subtracting forest cover areas from total geographical areas.

3.2.3 Indicators and Filtering the Samples

To carry out regression analysis, Land Degradation Index (LDI) is used as a dependent variable, and unforested land is used as an independent indicator. For regression of this model, 104 districts, firstly, are selected by few criteria as given following in Fig. 3.1.

3.3 Results and Discussion

Demographic pressure is being considered in India; hence land-man ratio has consistently declined. Due to rising athropogenic activity as a reult of population pressure, the decrease in forest cover in general are emerged as a new problem. The lessening of forest cover has a direct effect on degrading the quality of land (Lamb 2011). That is why there is a need to probe into the relationship among population, non-forested area, and land degradation. For further study, 104 districts, which are highly degraded, have been selected, and it is hypothesized that lack or loss of forest and population density directly affects land degradation. Non-forested land and population density have regressed on degraded land. The correlation among non-forested land, population density, and degraded land shows that there is positive relationship (i.e., 0.51). The f-value is 19.36, which is significant at the level of <0.001 (Table 3.1). T-value for non-forested land is 6.005 and for population density is −1.70, which are significant at the level of <0.001 and <0.092, respectively (Table 3.1). It reveals that if non-forested land grows and population density increases, the quality of land will be changed downward. It was intended that if population pressure on land will decrease and forestation on land will increase, the pace of degradation will be lowered (Table 3.2).

The outcome of correlation has shown that no doubt there are other agents responsible for land degradation, but deforestation and population pressure are

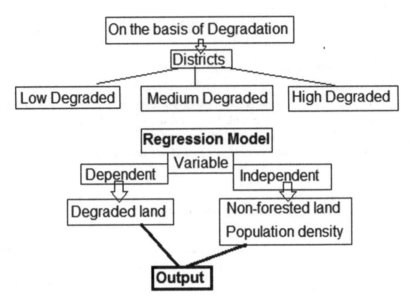

Fig. 3.1 Selection procedure

Table 3.1 Deforestation, population, and land degradation: model summary

R	R square	Adjusted R square	Std. error of the estimate	F
.516[a]	0.266692	0.252171	7.550357	18.37*

[a]Predictors: (constant), population density, non-forest area
*Significant at the level of <0.001

Table 3.2 Deforestation, population, and land degradation: coefficients

	Unstandardized coefficients		Standardized coefficients		
	B	Std. error	Beta	t	Sig.
(Constant)	1.7169	1.77		0.969	0.335
Non-forest (%)	0.17	0.03	0.517	6.005	0.000
Population density	−0.002	0.001	−0.146	−1.70	0.092

major factors for increment in degradation. The regression shows that it is significant at <0.001 level. The null hypothesis is rejected so that the increases in non-forested lands and population have high impact on degraded lands. The interesting finding is that forest degradation is not only crucial for human health but also for land health.

   The scarcity of natural resources is increasing with the rise in population, resultant, this leads to reduction of existing resources or degradation faster than they are being renewed naturally. This type of environmental and land degradation facilitates people with high income, risk, and health side effects. Increase in degradation in environment is mostly due to overpopulation (Premchander et al. 2003) and rural-urban migration (Shandra et al. 2003). Palm et al. (2011) have analysed how the environment and economy can interrelated at local and regional level. The plantation has not been proven feasible solution for sustainable use of natural resources; interestingly, low land productivity, and water scarcity like problems directly effect the plantation, which indirecly hampers the process for restoration of natural resources. Sahu and Dash (2011) try to explain the impact of mining activities on land degradation. Moreover, mining activities influence water, soil quality, and vegetation, including forest systems and human health. Mining activity exerted a long-lasting impact on landscape, ecosystem, and socio-cultural-economic considerations. Mining has been found to degrade the land to a significant extent. Overburden removal from the mining areas results in a very significant loss of rainforest and the rich top soil. The process of excavators and blasting normally does overburden removal; as a result, large volume of waste is the outcome of this. Topographic reconstruction, replacement of topsoil, and soil reconstruction are some reclamation strategies, which are supposed to be done in mining areas to mitigate the land degradation. Rehabilitation is also a good process to restore the areas or resources affected by mining. Extraction of natural resources like mining also affects land degradation. The above study is helpful in the understanding of one parameter of land degradation. The digging of rocks by which blasting has occurred in hilly areas.
   The analysis has emphatically evidenced that Indian land is severally suffering from gullied/ravine, scrublands and salinity, and alkalinity. As it appeared that there are four zones of land degradation, these zones are not only highly vulnerable due to human activities (due to deforestation or unplanned irrigation and agriculture or mining activities) (Menon and Bawa 1998; Gerold 2010; Sahu and Dash 2011; Pani et al. 2011), but these zones come under agro-ecological zone. The climatic variables (Clarke and Rendell 2007) and landscapes like river catchment etc. have remained significant factor of land degradation (Anthopoulou et al. 2006; Pani and Carling 2013; Morgan 2006). Without any doubt, the results stand as a proof of direct impact of deforestation on land degradation (Pimentel et al. 1986; Gupta et al. 1998; Reddy 2003; Turnbull et al. 2014). This evidence also suggested that if negative changes in forestation arise out in large scale, it may result in environmental problem (Jolly 1993; Hong and Ju 2007), resources problem (Natarajan et al. 2010), and food insecurity (Tekle 1999; Reddy 2003; Lal 2009; Bhattacharya and Guleria 2012; Parsons 2014). Most significantly, this study brings out a very open relationship among land degradation with non-forested cover and population pressure, and data reveals that degradation of land takes place with the decreasing forest cover land and increasing population density. To sum up, the need of strategy controlling deforestation, motivation to afforestation, and population control must be implemented to stop land degradation before it brings physical or financial ruin (Fig. 3.2).

Fig. 3.2 Aerial glimpse of Gangetic Plain land use, which is dominated by intensive agricultural activities. Red colour shows meandering, Yellow box represents human settlement, other green patches are either agricultural field or tree landscape

Regarding population growth, sustainability, and land degradation, Daily and Ehrlich (1992) stress that the major impact of population is a fraction of Earth's net primary productivity. The carrying capacity of the Earth is not like earlier because of current technological development, level of consumption, and socio-economic organization. For the sustainable use of resources with this development, there is a need to establish relevant temporal and spatial scales. Mortimore (1993) critically looks at the linkage between the growth of population in dryland and process of ecological degradation (desertification) in the developing world. In a Malthusian point of view, population growth results in increasing demand for food, and increasing cultivation areas are consequences of this. Due to spreading the cultivated land, there is reduction in length of fallows. Decline in soil fertility is an outcome of maximum use of land, and this results in declining yields, soil degradation, erosion, falling output environment destruction, and food scarcity. In contrast to a Malthusian view, a Boserupian view states that growth of population density even results in increasing demand for food and cash crops, but there is occurrence of increasing labour inputs per hectare. Adoption or diffusion of yield-enhancing technologies results in improvement in soil productivity, and a consequence of this is improvement in the yield per hectare. Due to this, there is increase in total production, land value, and investment in conservation of vegetation and productivity. One another study of Chopra and Gulati (1997) studied the linkage between environmental degradation (deforestation and land degradation) and movement of population from one region to another within developing countries. The hypothesis indicates that in the arid and semi-arid regions of India, a large part of outmigration is due to push factor such as environmental degradation and avoiding action of common property resources. Micro-experiment shows that the outmigration process becomes slow once the property rights are well defined, either in form of ownership or user rights (Fig. 3.3).

The understanding of current theories on population changes and the environment and land degradation is also essential before making or coming on any appropriate decision. In this context, two researches of Jolly (1993, 1994) evaluate the four theories on population, policies, economics, and land degradation categorized as neoclassical economists, classical economics and natural science perspectives, dependency and regional political ecology perspectives, and population as an intermediate variable. According to neoclassical economist, population growth is a neutral factor; it has no essential effect on land degradation. Degradation can occur when markets are not working efficiently. It may be short-term response to population growth. Land degradation can be the result of efficient depletion of land resources for production. Totally disagreeing with the first one, for the classical economists or natural scientist, high population growth is the independent factor causing land degradation. As an increasing population puts pressure on fixed available resources for maintaining or increasing the quality of livings, environmental degradation and land degradation occur as resources are depleting. Dependency theory, appearing similar, has stated that high population growth is a characteristic of a deeper problem, poverty. Land degradation and high population growth are deeply linked, not in that case when one causes the other, but in that case when their

Fig. 3.3 Photo shows how the plant roots protect soil from getting eroded or slow the rate of erosion. (Photo taken at Mirik, Darjeeling, West Bengal)

root cause is the same: unequal distribution of resources maintained by political and economic institutions. Analyst or last theory states that land degradation is ultimately the result of multiple factors. It is an exacerbating factor. This theory states that land degradation is a result of high population in combination with lack of employment opportunities and misconceived agricultural policies. The paper concludes that there is no consensus among the four principal frameworks. Most debate are centred on the two rival policies of the neoclassical economist and the classical economists or natural scientists.

Concerning the food security with land and environment issues, agricultural land reported increase in food output from an expansion of the area under cultivation as well as by using chemical, fertilizers, and irrigation. Food insecurity increases due to famine, conversion of cropland into non-farm use, thinning of topsoil, loss of irrigated land (Brown 1981), erosion and degradation of soil, decrease in biotic diversity (Ehrlich et al. 1993), increase in population, decline in per capita soil and water resources, increase in soil degradation and water pollution, and decrease in farm income (Lal 2011). Soil degradation affects human nutrition and health through its adverse impact on quantity and quality of food production. Soil degradation reduces crop areas by increasing harm to drought stress and elemental imbalances (Lal 2009).

There is a little concern to all human beings on how land degradation affects the basics of human life. The above literature gives us an idea about interlink among soil erosion, land degradation, food security, and poverty. Land degradation is one of the factors, which influence poverty, so it is necessary to understand how much human beings are affected by soil erosion, low productivity, and land degradation. Poverty comes under the social indicator of region. It is necessary to understand the social factors as well as physical factors.

With increasing pressure on land, population growth, and slow growth rate of economy, land degradation has become a major problem in the country. Land degradation factors identified are soil erosion and sedimentation, soil fertility decline, acidification, increase in salinity and alkalinity, accumulation of toxic substances, eutrophication due to overuse of fertilizers, leaching of groundwater, iron toxicity, pollution, and soil compaction. Loss of land productivity due to land degradation is US$ 36/ha/yr., while the loss due to nutrient depletion is about US$ 51/ha/yr. Lack of knowledge and less information availability on soil erosion have been found. A database should develop in these lines – fertility decline, salinity, and eutrophication – and recommendations for arresting land degradation have been highlighted in the paper of Mapa (2011).

3.4 Conclusion

Land in India is suffering from different types of land degradation such as gullied/ravine, scrublands, salinity, and alkalinity. At present, there are about 130 million ha of degraded land in India. Approximately, 28% of it belongs to the category of forest-degraded area, 56% of it belongs to water-eroded area, and the rest is affected by saline and alkaline activities. The second chapter highlights the four zones of high land degradation. The high degraded zones – Thar zone, south Deccan zone, north Deccan zone, and northeastern zone – are not only results of human activities but also by climatic factors like rainfall and landscapes, like river catchment, etc. Without a doubt, the results stand as a proof of the direct impact of deforestation on land degradation. The result has also suggested that if increment in non-forested land is arising out in large scale, it might result not only in environmental problem but also resource problem and food insecurity. On the ground reality, on one hand the deforestation occurred mostly due to increment in agricultural land in purpose of boosting agricultural productivity. However, on the other hand, most significantly, this study brings out a very open relationship among land degradation with non-forested area and population pressure, and results uncover that degradation of land takes place with the decreasing forest cover land and increasing population density.

References

Anthopoulou, B., Panagopoulos, A., & Karyotis, Th. (2006). The Impact of Landscape in Northern Greece. Landslides, 3, 289–294. https://doi.org/10.1007/s10346-006-0056-x.

Bhattacharya, T & Guleria, S. (2012). Costal Flood Management in Rural Planning Unit Through Land-Use Planning: West Bengal, India. Journal of Coastal Conservation, 16, 77–87. https://doi.org/10.1007/s11852-011-0176-x.

Brown, L. R. (1981). World Population Growth, Soil Erosion and Food Security. Science, New Series, 214 (4524), 995–1002. Retrieved from http://www.jstor.org/stable/1686685

Chopra, K. & Gulati, S.C. (1997). Environmental Degradation and Population Movements: The Role of Property Rights. Environmental and Research Economics, 9, 383–408.

Clarke, M. L. & Rendell, H. M. (2007). Climate, Extreme Event and Land Degradation. In Mannava V. K. Sivakumar & Ndegwa Ndiang'ui (Eds.), Climate and Land Degradation, (pp.137–149). NY, Springer Berlin Heidelberg.

Daily, G. C. & Ehrlich, P. R. (1992). Population, Sustainability and Earth's carrying capacity. BioScience, 42 (10), 761–771. Retrieved from http://www.jstor.org/stable/1311995

Ehrlich, P. R., Ehrlich, A. H. & Daily, G. C. (1993). Food Security, Population and Environment. Population and Development Review, 19 (1), 1–32. Retrieved from http://www.jstor.org/stable/2938383

Jolly, C. L. (1993). Population Change, Land Use and the Environment. Reproductive Health Matters, 1, 13–25. Retrieved from http://www.jstor.org/stable/3774852

Jolly, C. L. (1994). Four Theories of Population Change and the Environment. Population and Environment, 16 (1), 61–90. Retrieved from http://www.jstor.org/stable/27503376

Gerold, G. (2010). Soil and Water Degradation Following Forest Conversion in Humid Tropics (Indonesia). In P. Zdruli, M. Pagliai, S. Kapur, & A. F. Cano (Eds.), Land Degradation and Desertification: Assessment, Mitigation and Remediation, (pp.267–284). London, NY, Springer Dordrecht Heidelberg.

Gupta, S. K., Ahmed, M., Hussain, M., Pandey, A. S., Singh, P., Saini, K. M…Das, S.N. (1998). Inventory of Degraded Lands of Palamau District, Bihar- A Remote Sensing Approach. Journal of the Indian Society of Remote Sensing, 26 (4), 161–168.

Hong, M. & Ju, H. (2007). Status and Trends in Land degradation in Asia. In Mannava V. K. Sivakumar & Ndegwa Ndiang'ui (Eds.), Climate and Land Degradation, (pp.55–64). NY, Springer Berlin Heidelberg.

Lal, R. (2009). Soil Degradation as a Reason for Inadequate Human Nutrition. Food Security, 1, 45–57. https://doi.org/10.1007/s12571-009-0009-z.

Lal, R. (2011). Soil Degradation and Food Security in South Asia. In Climate Change and Food Security in South Asia. https://doi.org/10.1007/978-90-481-9516-9_10

Lamb, D. (2011). Forest and Land Degradation in the Asia-Pacific Region. In Regreening the Bare Hills, World Forest 8, 41–91. https://doi.org/10.1007/978-90-481-9870-2_2.

Mapa, R. B. (2011). Strategies for Arresting Land Degradation in South Asian Countries-Sri Lankan Experience. In Dipak Sarkar, A. B. Azad, S. K. Singh & N. Akter (Eds.), Strategy for Arresting Land Degradation in South Asian Countries, (pp. 151–170). Dhaka, Bangladesh, SAARC Agriculture Centre.

Mortimore, M. (1993). Population Growth and Land Degradation. GeoJournal, 31 (1), 15–21. Retrieved from http://www.jstor.org/stable/41145902

Menon, S. & Bawa, K. S. (1998). Deforestation in the Tropics: Reconciling Disparities in Estimates for India. Amibio, 27 (7), 576–577. Retrieved from http://www.jstor.org/stable/4314794

Morgan, R. P.C. (2006). Managing Sediment in the Landscape: Current Practices and Future Vision. In P. N. Owens (Eds.), Soil Erosion and Sediment Redistribution in River Catchments, (pp.287–293). Oxfordshire, UK, Biddles Ltd, King's Lynn.

Natarajan, A., Janakiraman, M., Manoharan, S., Kumar, K. S. A., Vadivelu, S. & Sarkar, D. (2010). Assessment of Land Degradation and Its Impacts on Land Resources of Sivagangai Block, Tamil Nadu, India. In P. Zdruli, M. Pagliai, S. Kapur, & A. F. Cano (Eds.), Land Degradation and Desertification: Assessment, Mitigation and Remediation, (pp. 235–252). London, NY, Springer Dordrecht Heidelberg.

Palm, M., Ostwald, M., Murthy, I. K., Chaturvedi, R. K. & Ravindranath, N. H. (2011). Barriers to Plantation Activities in Different Agro-Ecological Zones of Southern India. Regional Environmental Change, 11, 423–435. https://doi.org/10.1007/s10113-010-0154-0

Pani, P., Mishra, D. K. & Mohapatra, S. N. (2011). Land Degradation and Livelihoods in Semiarid India: A Study of Formers' Perception in Chambal Valley. Asian Profile, 39 (5), 505–519.

Pani, P. & Carling, P. (2013). Land Degradation and Spatial Vulnerabilities: A Study of Inter-Village Differences in Chambal Valley, India. Asian Geographer, 30 (1), 65–79. Retrieved from https://doi.org/10.1080/10225706.2012.754775

Parsons, A. J. (2014). Abandonment of Agricultural Land, Agricultural Policy and Land Degradation in Mediterranean Europe. In E. N. Mueller, J. Wainwright, A. J. Parsons & L. Turnbull (Eds.), Patterns of Land Degradation in Drylands: Understanding Self-Organised Ecogeomorphic Systems, (pp.357–366). London, NY, Springer Berlin Heidelberg.

Pimentel, D., Dazhong, W., Sanford, E., Lang, H., Emerson, D. & Karasik, M. (1986). Deforestation: Interdependency of Fuelwood and Agriculture. Oikos, 46 (3), 404–412. Retrieved from http://www.jstor.org/stable/3565841

Premchander, S., Jeyaseelan, L. & Chidambaranathan, M. (2003). In Search of Water in Karnataka, India: Degradation of Natural Resources and Livelihood Crisis in Koppal District. Mountain Research and Development, 23 (1), 19–23. Retrieved from http://www.jstor.org/stable/3674530

Priya, R. (2014). Extent, Pattern and Implication of Land Degradation in India: Analysis of District-wise variation. Unpublished dissertation submitted at the Jawaharlal Nehru University, New Delhi.

Priya, R. and Pani, P. (2015). Land Degradation and Deforestation in India: A District Level Analysis. Annals of National Association of Geographer, India, XXXV (1), June, 50–70. ISSN: 0970-972X

Rao, M. M. (2009). Linear Regression for Random Measures. In Ashis SenGupta (Ed.), Advances in Multivariate Statistical Methods, (pp. 131–145). Singapore, World Scientific Publishing Co. Pte. Ltd.

Reddy, V. R. (2003). Land Degradation in India: extent, costs and Determinant. Economic Political Weekly, 38 (44), 4700–4713. Retrieved from http://www.jstor.org/stable/4414225

Sahu, H. B. & Dash, S. (2011). Land Degradation due to Mining in India and its Mitigation Measures. 2nd International Conference on Environmental Science and technology, IPCBEE, 6, 132–136. Singapore, IACSIT Press.

Shandra, J. M., London, B. & Williamson J. B. (2003). Environmental Degradation, Environmental Sustainability, and Overurbanization [Over-urbanization] in the Developing World: A Quantitative, Cross-National Analysis. Sociological Perspectives, 46 (3), 309–329. Retrieved from http://www.jstor.org/stable/10.1525/sop.2003.46.3.309

Tekle, K. (1999). Land Degradation Problem and Their Implications for Food Shortage in South Wello, Ethiopua. Environmental Management, 23 (4), 419–427.

Turnbull, L., Wainwright, J. & Ravi, S. (2014). Vegetation Change in the Southwestern USA: Pattern and Process. In E. N. Mueller, J. Wainwright, A. J. Parsons & L. Turnbull (Eds.), Patterns of Land Degradation in Drylands: Understanding Self-Organised Ecogeomorphic Systems, (pp.289–314). London, NY, Springer Berlin Heidelberg.

Chapter 4
Land Degradation in India and Propinquity with Rainfall

Abstract In tropical areas, poor soil structure, high potential capacity of rainfall, and wind directly cause soil erosion. In developing countries, due to lack of adequate pesticides and climatic variability, crop yields of verious crops have been impacted. The poor farmers of developing countries could not afford to deal with fluctuating climate due to incapability. This chapter will enable for a deeper understanding towards the rainfall-land degradation relation. This chapter has first briefed about the climatic status in India spatially. Thereafter, relation between the two (land degradation and rainfall) was tried to understand by using the method of regression analysis. It is observed that rainfalls received by Indian subcontinent are high during summer monsoon, while arid and semi-arid region rainfall obtained in June, July, August, and September and remaining months has dried up rapidly. Both the factors responsible for land degradation, for instance, heavy and splash rainfall, cause high soil erosion, and lack of rainfall does not support the vegetation improvement and survival.

Keywords Land degradation · Wasteland · Climate · Rainfall · Regression analysis

4.1 Introduction: Climate and Land Degradation

"A likely story is man brought up the land degradation, ergo mostly defined as "the prime maker of desert" (Kishk 1990), while environmental aspects also influence directly to land degradation, and it mostly covers arid and semi-arid regions, resultant of climatic variation and human activities (UNCCD 1994). The concept Zhang et al. (2011), who worked on Shiyanghe Basin, has analysed land degradation by using precipitation, temperature, and solar radiation data and land use data to estimate the value of net primary productivity in the area. The study has concluded that the degradation in basin areas was caused by almost 80% human interjection while the rest by climatic mutation. However, the study has supported that vegetation restoration occurred or emerged both due to human intervention and climate changen controlling land degradation (Kishk 1990, Zhang et al. 2011). Though, precipitation is not helpful to estimate the impact on land degradation in totality, but it is aide to insinuate whether the intensified land degradation problem or not. Different geographical areas suffer severely with land degradation problem in India; in addition to human-induced factors, many components of climatic condition like temperature, humidity, evapotranspiration, wind speed, and direction too play major role to become important determinant of it. Meadows and Hoffman (2003) concluded in their study that both the greater rainfall and increased air temperature progress the evapotranspiration and influence lower soil moisture availability. These climatic variations are able to change the condition of land degradation. Qiang et al. (2011) estimated that the Tropical Rainfall Measuring Mission (TRMM) data based on rainfall intensity significantly correlated with the interpolated rain gauge data and impacts on soil erosion. This research strongly has breathed that rainfall is the main external factor contributing to erosion of soil by water (Qiang et al. 2011).

In tropical areas, poor soil structure, high potential capacity of rainfall, and wind directly cause soil erosion. In developing countries, due to lack of adequate pesticides and climatic variability, crop yields of verious crops have been impacted. The poor farmers of developing countries could not afford to deal with fluctuating climate due to incapability (Grepperud 1997). In coastal condition, especially reference to monsoon regions, temperature is not the only factor to increase rainfall, but the sea surface and the wind interaction with land surface also accelerate the rainfall (Paeth and Friederichs 2004).

4.2 Methodology and Database

To understand, as per literature concern, rainfall is one of the factors which affect and increase the pace of the land degradation, and the following methods are used.

4.2.1 Statistical Methods

Regression will be calculated to observe the result for finding the third objective. Here it is supposed that rainfall of monsoon season is directly correlated to land degradation. If there is increase in rainfall, then land degradation also increases. It means there is positive relation between land degradation and rainfall. In regression model, the following equation has been run in SPSS software for all degraded zones in India:

$$LDI = \beta 0 + \beta 1 \ X1 + \beta 2 \ X2 + \beta 3 \ X3 + \beta 4 \ X4 + u$$

Here, LDI is Land Degradation Index, X1 is monthly rainfall of June, X2 in rainfall in July, X3 is rainfall in August, and X4 is rainfall in September.

4.2.2 Source of Data

Land degradation and wasteland data have been taken from NRSC. Rainfall data have been taken from Indian Meteorological Department, Govt. of India.

4.2.3 Indicators and Filtering the Samples

Regression model has been done separately for all the highly degraded zones, which accurately comes under the agro-ecological zones (division by Indian Agricultural Statistics Research Institute) having different rainfall characteristics. Sixteen districts are taken from Thar zone, 13 districts from North Deccan zone, 13 districts from south Deccan zone, and 7 districts from the northeastern region of India. The plan for model is given below (Fig. 4.1).

4.3 Climatic Condition in India

India has extraordinary varieties not only in climate but also in the physiographic, geological evaluation, vegetation cover, etc., all of them mostly and hardly affect land degradation. However, only rainfall is focused in the study. Thus, the description about rainfall condition in India has been given, which varied due to its position on globe. India undergoes yearly rainfall resulting from the reversal of winds from January to July. Extreme heat of summer months results in the landmass of the Indian subcontinent to become hot and attract the moist air from the Indian Ocean. It is causing the reversal of the winds over the regions called the summer or southwest monsoon. Cosequently, most of rainfall in India (above 75%) is received during a short span of 4 months (June–September). The received rainfall during monsoon season influences agriculture, water resources, power generation, economics, and ecosystem in the county of the greatest intensity. The minimum average rainfall is less than 13 cm over the western rainfall; however, the Mawsynram in the Meghalaya has received as much of 1141 cm.

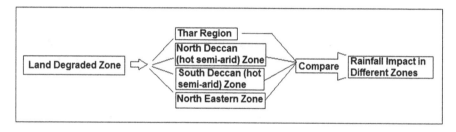

Fig. 4.1 Process of analysis

4.3.1   Winter Season (January and February)

According to Indian Meteorological Department (GOI), this season starts in early December associated with clear skies, fine weather, light northerly winds, low humidity and temperature, and large daytime variations of temperature. The rains during this season generally occur over the western Himalayas, the extreme northeastern parts, Tamil Nadu, and Kerala. Western disturbances and associated trough in westerlies are the main rainfall-bearing system in the northern and eastern parts of the country.

4.3.2   Pre-monsoon Season (March, April, and May)

The temperature begins to change and is slightly increased in March, and this season also experiences cyclonic storms, which have intense low pressure system over hundreds to thousands of kilometres associated with the surface wind more than 33 knots over the Indian seas, viz., Bay of Bengal and the Arabian Sea. These systems generally move towards a northwesterly direction, and some of them recurve to northerly or northeasterly path. Storms forming over the Bay of Bengal are more frequent than the ones originating over the Arabian Sea. On average, the frequency of storms is about 2.3 per year. Local severe storms or violent thunderstorms associated with strong winds and rains lasting for short durations occur over the eastern and northeastern parts over Bihar, West Bengal, and Assam. This is called "northwesters" or "Kalbaisakhis" locally.

4.3.3   Summer Monsoon (June, July, August, and September)

The southwest monsoon is the most significant feature of Indian climate. This season is spread over 4 months, but the actual period at a particular place depends on the onset and withdrawal dates. It varies from less than 75 days in the west of Rajasthan to more than 120 days over the southwestern region of the country contributing to about 75% of the annual rainfall. Southwest monsoon current becomes feeble and generally starts withdrawing from Rajasthan by September 1 and from the northwestern parts of India by September 15. It withdraws from almost all parts of the country by October 15 and is replaced by a northerly continental airflow called northeast monsoon. The retreating monsoon winds cause occasional showers along the east coast of Tamil Nadu, but rainfall decreases away from coastal regions.

4.3.4   Post-monsoon (October, November, and December)

Northeast (NE) monsoon or post-monsoon season is transition season associated with the establishment of the northeasterly wind regime over the Indian subcontinent. Meteorological subdivisions, namely, Coastal Andhra Pradesh Rayalaseema, Tamil Nadu, Kerala, and South Interior Karnataka, receive good amount of rainfall accounting for about 35% of their annual total in these months. Many parts of Tamil Nadu and some parts of Andhra Pradesh and Karnataka receive rainfall during this season due to the storms forming in the Bay of Bengal.

Large-scale losses to life and property occur due to heavy rainfall, strong winds, and storm surge in the coastal regions. The day temperatures start falling sharply all over the country. The mean temperature over the northwestern parts of the country is declining from about 38 °C in October to 28 °C in November. Decrease in humidity levels and clear skies over most parts of north and central India after mid-October are characteristic features of this season.

4.4   Land Degradation and Rainfall: Indian Status

The rainfall in India at average (actual) was 1026.9 mm; in 2008–2009, it reached at 1936.21 mm (Figs. 4.2 and 4.3). Appearance of actual average rainfall has crept in the duration of 1996–2008. Based on temporal and spatial distribution of rainfall, the country is divided into six homogenous rainfall zones based on homogenous rainfall characteristics (Ramachandran and Kedia 2013) (Fig. 4.4).

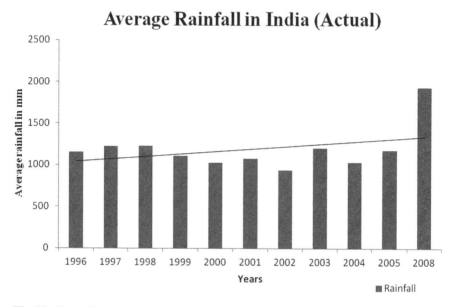

Fig. 4.2 Trend of average annual rainfall from 1996 to 2008

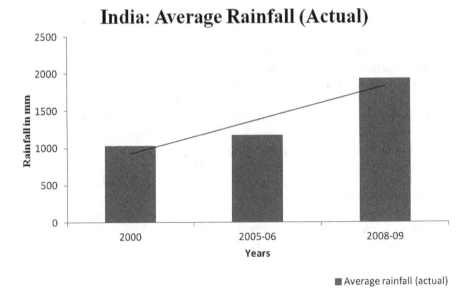

Fig. 4.3 Trend of average rainfall in India (actual)

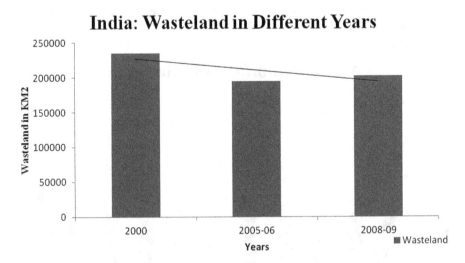

Fig. 4.4 Trend of degraded land in India

4.4.1 Rainfall in Winter Season

Jammu and Kashmir receives high actual rainfall of 238.7 mm, and similarly it gains high normal rainfall in winter season through western disturbances (Ramachandran, and Kedia 2013) followed by Himanchal Pradesh, Uttarakhand, Arunachal Pradesh, and West Bengal (Figs. 4.5 and 4.6). The detailed description of rainfall is given in Tables 4.1 and 4.2.

4.4.2 Rainfall in Pre-monsoon Season

Whereas in the pre-monsoon season, the transformed condition of the Indian subcontinent causes Assam, Meghalaya, Arunachal Pradesh, and Western Ghats to obtain very high actual rainfall in 2008. High rainfall of 165.5 mm is carried out by West Bengal, Tamil Nadu, Karnataka, Nagaland, Manipur, Tripura (Tables 4.3 and 4.4), while the highest normal rainfall is received by Assam, Meghalaya, Manipur, Tripura, Sikkim, Nagaland, Jammu and Kashmir, and Kerala in 2008 (Figs. 4.7 and 4.8).

4.4.3 Rainfall in Monsoon Season

In monsoon season, a thick rainfall is received due to proceeding southwest monsoon towards the Himalayan region (Murata et al. 2007). Compared to other seasons, the Indian subcontinent received highest rainfall in the said season. Thus, the distribution of monsoon rainfall is very important to correlate with land degradation. The Indian monsoon is a complex phenomenon that encompasses a vast variety of spatial and temporal scales and comprises interaction between land and ocean and atmosphere (Ramachandran & Kedia 2013). Fig. 4.9 illustrates that Assam, Meghalaya, Arunachal Pradesh, Sikkim, and West Bengal gain more than 1500 mm normal rainfall in 2008 (Table 4.5). Similarly, it demonstrates that states like Arunachal Pradesh, Sikkim, West Bengal, Assam, Meghalaya, and Kerala attain highest rainfall during monsoon season, having more than 1500 mm annual actual rainfall followed by Orissa, Maharashtra, and Karnataka receiving more than 1200 mm actual annual rainfall (IMD, Pune, GoI). In Maharashtra, the rainfall concentration occurs over the Konkan region (Table 4.6). With the northward propagation of the Arabian Sea branch of the monsoon during July and August, there is increase in rainy days over the Konkan region (Ratna 2012). In monsoon season, Rajasthan, Jammu and Kashmir, Haryana Tamil Nadu, and Punjab accomplish very low normal rainfall (less than 600 mm); while low actual rainfall in monsoon season is found over Haryana, Jammu and Kashmir, Rajasthan, and Tamil Nadu (less than 600 mm), and Punjab has received 603.8 mm actual rainfall in 2008 (Fig. 4.10). Rajasthan, which do not receive high rainfall, goes through drought many times (Jain et al. 2010).

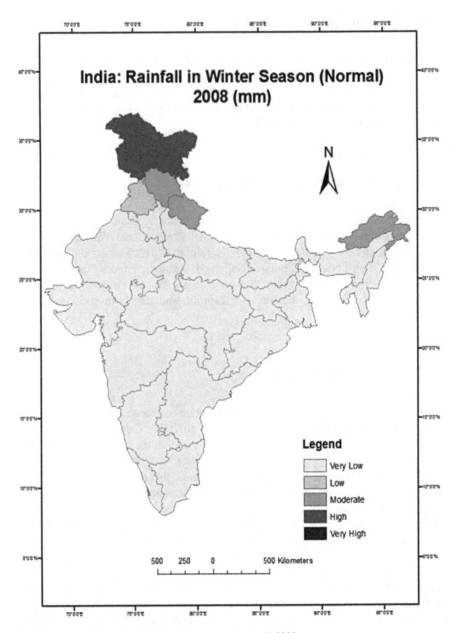

Fig. 4.5 Distribution of rainfall in winter season (normal) 2008

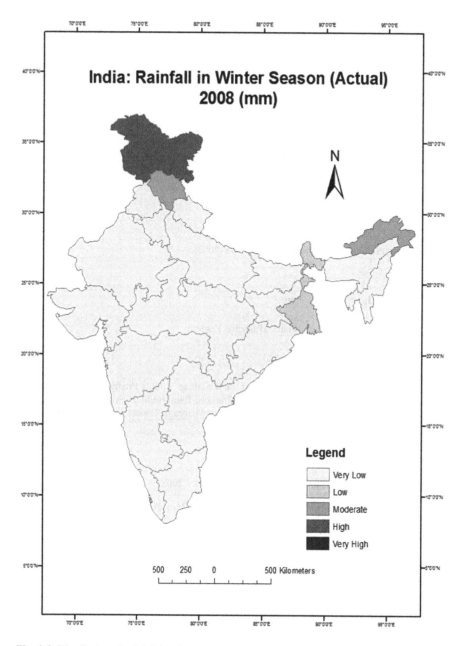

Fig. 4.6 Distribution of rainfall in winter season (actual) 2008

Table 4.1   Rainfall (normal) in India: winter season (2008)

Very high (>200 mm)	NA
High (150–200 mm)	Jammu and Kashmir
Moderate (100–150 mm)	Uttaranchal, Arunachal Pradesh, and Himachal Pradesh
Low (50–100 mm)	Punjab
Very low (<50 mm)	Gujarat, Karnataka, Maharashtra, Rajasthan, Andhra Pradesh, Kerala, Bihar, Madhya Pradesh, Orissa, Uttar Pradesh, Tamil Nadu, Haryana, Chandigarh, Delhi, Jharkhand, Sikkim, West Bengal, Nagaland Manipur Mizoram, Tripura, Assam, and Meghalaya

Table 4.2   Rainfall (actual) in India: winter season (2008)

Very high (>200 mm)	NA
High (150–200 mm)	Jammu and Kashmir
Moderate (100–150 mm)	Arunachal Pradesh and Himachal Pradesh
Low (50–100 mm)	Sikkim and West Bengal
Very low (<50)	Gujarat, Karnataka, Maharashtra, Rajasthan, Andhra Pradesh, Kerala, Bihar, Madhya Pradesh, Orissa, Uttar Pradesh, Tamil Nadu, Haryana, Chandigarh, Delhi, Jharkhand, Nagaland Manipur Mizoram, Tripura, Assam, Meghalaya, Uttaranchal, and Punjab

Table 4.3   Rainfall (normal) in India: pre-monsoon season (2008)

Very high (>400 mm)	Arunachal Pradesh, Assam, Meghalaya, Nagaland Manipur Mizoram, Tripura, Kerala, Jammu and Kashmir
High (150–400 mm)	Sikkim, West Bengal, and Himachal Pradesh
Moderate (100–150 mm)	Orissa, Uttaranchal, Tamil Nadu, and Karnataka
Low (50–100 mm)	Bihar, Andhra Pradesh, Jharkhand, and Punjab
Very low (<50 mm)	Gujarat, Maharashtra, Rajasthan, Madhya Pradesh, Uttar Pradesh, Haryana, Chandigarh, and Delhi

Table 4.4 Rainfall (actual) India: pre-monsoon season (2008)

Very high (>400 mm)	Arunachal Pradesh, Assam, Meghalaya, Kerala
High (150–400 mm)	Sikkim, West Bengal, Nagaland Manipur Mizoram, Tripura, Jammu and Kashmir, Tamil Nadu, Karnataka
Moderate (100–150 mm)	Orissa, Uttaranchal, Andhra Pradesh, Himachal Pradesh
Low (50–100 mm)	Bihar, Jharkhand and Punjab, Haryana, Chandigarh, Delhi
Very low (<50 mm)	Gujarat, Maharashtra, Rajasthan, Madhya Pradesh, Uttar Pradesh

4.4.4 Rainfall in Post-monsoon Season

Monsoon leaves in October, and post-monsoon period starts. In 2008, Kerala and Tamil Nadu have collected high normal rainfall in post-monsoon period followed by Arunachal Pradesh having high normal rainfall. Gujarat, Rajasthan, Haryana, and Punjab (Table 4.7) have gained very low normal rainfall in 2008 (Fig. 4.11). Except Kerala and Tamil Nadu, all states have received either moderate rainfall, low rainfall, or very low actual rainfall. Thar Desert, Middle India, and Gangetic plain except West Bengal (Table 4.8) had very low actual rainfall (Fig. 4.12).

4.4.5 Annual Average Rainfall

Due to location at the monsoon zone, coastal South Indian states and northeastern states derive high average rainfall. In 2008 Kerala had the highest annual normal rainfall followed by Arunachal Pradesh, Assam, Meghalaya, Sikkim, West Bengal, Nagaland, Manipur, Mizoram Karnataka, and Uttar Pradesh (Table 4.9), having more than 1500 mm annual normal rainfall (Fig. 4.13).

Similarly, highest annual rainfall (actual) is received by Kerala, Arunachal Pradesh, Assam, Meghalaya, Sikkim, West Bengal, and Karnataka, while Nagaland, Manipur, and Mizoram have received less than 1500 mm actual annual rainfall in 2008. Moreover, Maharashtra, Bihar, Uttarakhand, and Jharkhand have obtained high annual actual rainfall (Table 4.10), varying between 1000 and 1500 mm (Fig. 4.14).

A study by Doi (2001) on Rajasthan has explained that the arid and semi-arid region receive 70% of rainfall in a short period of time. Due to not receiving good rainfall, most of the area remained without vegetation cover or cropped (Jain et al. 2010). In contrary, Mawsynram in Meghalaya received highest rainfall in the world. The daily rainfall variation occurs due to firstly the effect of the synoptic scale disturbance with a periodicity of 10–20 days and another due to topography (Murata et al. 2007). A study by Francis and Gadgil (2006) concluded that the west coast of the Indian peninsula receives very heavy rainfall during summer monsoon, sometimes exceeding even 15 cm day^{-1} or 20 cm day^{-1}.

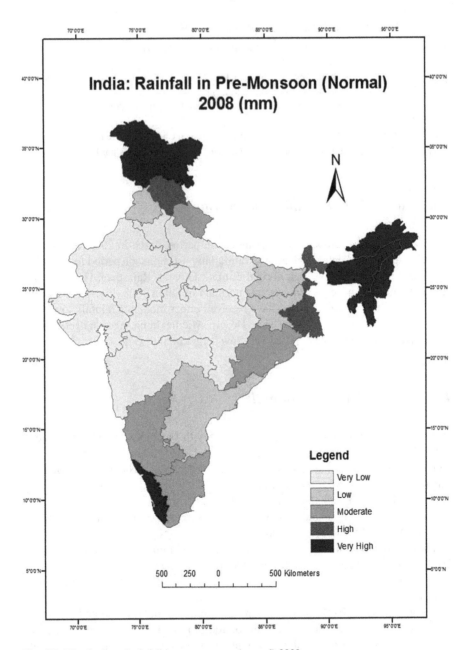

Fig. 4.7 Distribution of rainfall in pre-monsoon (normal) 2008

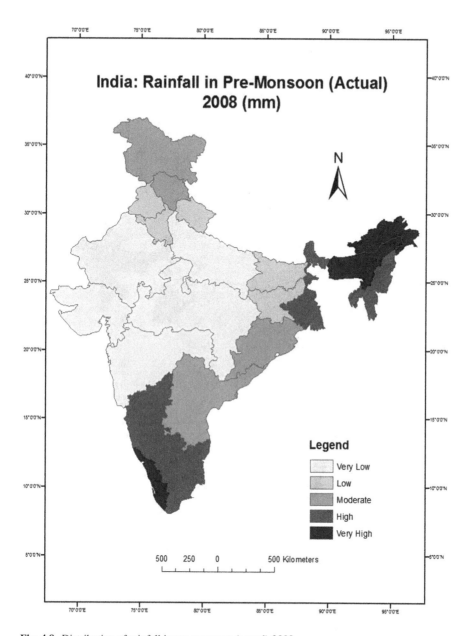

Fig. 4.8   Distribution of rainfall in pre-monsoon (actual) 2008

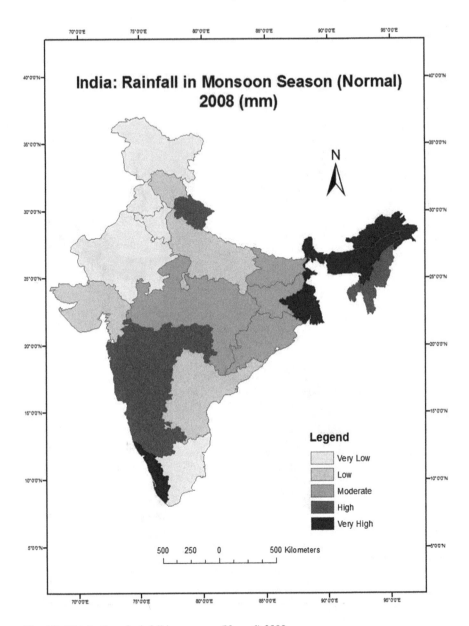

Fig. 4.9 Distribution of rainfall in monsoon (Normal) 2008

Table 4.5 Rainfall (normal) in India: monsoon season (2008)

Very high (>1500 mm)	Arunachal Pradesh, Assam, Meghalaya, Kerala, Sikkim, West Bengal
High (1200–1500 mm)	Nagaland Manipur Mizoram, Tripura, Karnataka, Maharashtra, Uttarakhand
Moderate (900–1200 mm)	Orissa, Madhya Pradesh, Bihar, Jharkhand
Low (600–900 mm)	Gujarat, Uttar Pradesh, Andhra Pradesh, Himachal Pradesh
Very low (<600 mm)	Rajasthan, Punjab, Haryana, Chandigarh, Delhi, Jammu and Kashmir, Tamil Nadu

Table 4.6 Rainfall (actual) in India: monsoon season (2008)

Very high (>1500 mm)	Arunachal Pradesh, Assam, Meghalaya, Kerala, Sikkim, West Bengal
High (1200–1500 mm)	Karnataka, Maharashtra, Orissa
Moderate (900–1200 mm)	Bihar, Jharkhand, Nagaland Manipur Mizoram, Tripura, Uttaranchal, Uttar Pradesh
Low (600–900 mm)	Gujarat, Andhra Pradesh, Himachal Pradesh, Madhya Pradesh, Punjab
Very low (<600 mm)	Rajasthan, Haryana, Chandigarh, Delhi, Jammu and Kashmir, Tamil Nadu

4.4.6 Land Degradation in Indian States

The data shows that Jammu and Kashmir have the highest degraded land of its total geographical areas having around 74.4% areas under wasteland followed by Himachal Pradesh with 40.1% areas under this category. Rajasthan, Uttar Pradesh, Nagaland, Manipur, Mizoram, Tripura, and Arunachal Pradesh are highly degraded having 24.8%, 24.0%, 23.9% (combined value of Nagaland, Manipur, Mizoram, and Tripura), and 17.8% wasteland areas. Haryana, Punjab, and Uttarakhand are showing lowly degraded land of total geographical land (Fig. 4.15).

Area-wise, Rajasthan comes first having 84929.1 km^2 areas under wasteland followed by Jammu and Kashmir (75435.8 km^2), Madhya Pradesh (40113.3 km^2), Maharashtra (37830.8 km^2), and Andhra Pradesh (37296.6 km^2). Again Haryana, Punjab, and Uttarakhand come in last (NRSC 2008).

4.5 Land Degradation and Rainfall

Land degradation and deforestation both are closely related to rainfall, and it is expressed as supposition that the land degradation induces changes in rainfall in the long term. While other than that, on shorter period up to inter-seasonal timescale, soil moisture is a key factor in the rainfall process (Paeth and Friederichs 2004). High rainfall is also responsive to high soil erosion, because deforestation increases the run-off and results into high erosion (Harden and Mathews 2000). Here, it is

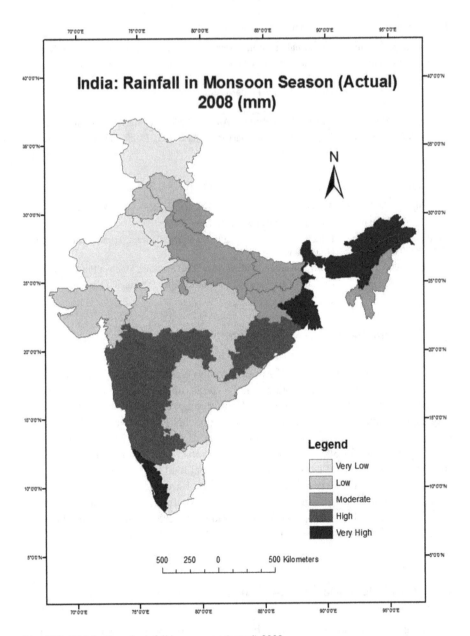

Fig. 4.10 Distribution of rainfall in monsoon (actual) 2008

Table 4.7 Rainfall (normal) in India: post-monsoon season (2008)

Very high (>400 mm)	Tamil Nadu, Kerala
High (200–400 mm)	Arunachal Pradesh
Moderate (100–200 mm)	Orissa, Andhra Pradesh, Himachal Pradesh, Jharkhand, Maharashtra, Sikkim, West Bengal, Nagaland Manipur Mizoram, Tripura, Jammu and Kashmir, Karnataka, Assam, Meghalaya
Low (50–100 mm)	Bihar, Madhya Pradesh, Uttar Pradesh, Uttaranchal
Very low (<50 mm)	Punjab, Haryana, Chandigarh, Delhi, Gujarat, Rajasthan

tried to know how rainfall[1] is correlated with land degradation in highly degraded districts[2].

To investigate this question, regression has been run by taking the Land Degradation Index as a dependent variable and monthly rainfall as independent variables. The R^2 is 0.65, which explains 65% relation between land degradation and rainfall. The f-statistics is significant at the level of 0.001 (Table 4.11). Literature suggests that the changes in land degradation might be upshot due to rainfall (Meshesha et al. 2014).

Table 4.12 explains rainfall of which months significantly affect land degradation. The t-values of February, March, May, June, July, September, October, November, and December are highly significant, while t-values of January, April, and August are significant at the level of 0.62, 0.92, and 0.91, respectively (Table 4.12). It has explicated that degradation of soil has taken place during monsoon.

4.6 Zone-Wise Analysis of Land Degradation and Rainfall

4.6.1 Thar Zone: Land Degradation and Rainfall

One other study on rangeland has drawn that drought is to be expected in arid and semi-arid region because rainfall ranges from one-tenth to twice the yearly average (Milton et al. 1994). Similarly, Meshesha et al. (2014) have worked on the central rift valley of Ethiopia that comes under arid and semi-arid climate. It has revealed that rainfall causes high rate of soil erosion and poor vegetation cover, and rainfall is determined by rain intensity, raindrop size, and kinetic energy. Nevertheless, this

---

[1] There are so many climatic factors, such as rainfall, temperature, wind direction, etc. But in this study, rainfall is taken to understand the relation with land degradation.

[2] All district-wise analysis is not done because rainfall data for all districts is not available and rain gauge station is not established in all the districts by the Indian Meteorological Department.

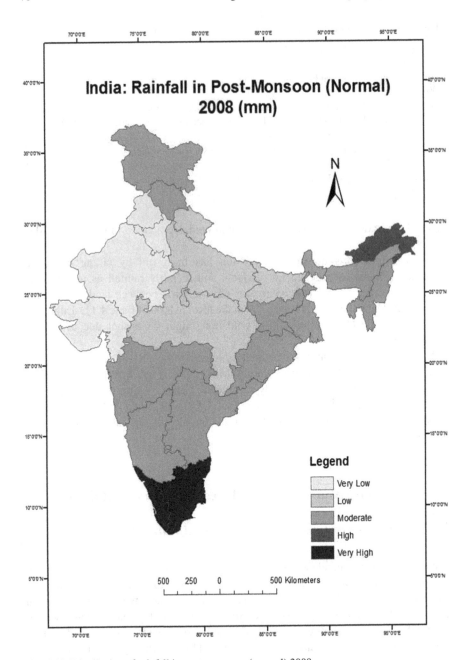

Fig. 4.11 Distribution of rainfall in post-monsoon (normal) 2008

Table 4.8 Rainfall (actual) in India: post-monsoon season (2008)

Very high (>400 mm)	Tamil Nadu, Kerala
High (200–400 mm)	NA
Moderate (100–200 mm)	Andhra Pradesh, Nagaland Manipur Mizoram, Tripura, Jammu and Kashmir, Karnataka, Assam, Meghalaya, Arunachal Pradesh
Low (50–100 mm)	Sikkim, West Bengal, Himachal Pradesh
Very low (<50 mm)	Punjab, Haryana, Chandigarh, Delhi, Gujarat, Rajasthan, Jharkhand, Bihar, Madhya Pradesh, Uttar Pradesh, Uttaranchal, Orissa, Maharashtra

research only addresses the spatial variation of monthly rainfall and impact on land degradation. In this region, 16 districts, which are highly degraded in this zone, are selected to assess the rainfall.

Jaisalmer is highly degraded district in India and Thar region is one of region with highest areas under land degradation (according to Land Degradation Index and area-wise also) and possess 29.69% areas under wasteland of the selected indicators of total geographical area. This region receives very low rainfall. Apart from this, it also attains comparatively very low rainfall in the monsoon season. Jaisalmer receives rainfall of 5.1 mm in June, 52.9 mm in July, 153.7 mm in August, and 11.3 mm in September (Table 4.2). The same condition occurs in all the districts of the region, namely, Kuchchh, Barmer, Rajkot, Pali, Rajsamand, Udaipur, Bhilwara, Ajmer, Jaipur, Surendranagar, Jamnagar, Bhavnagar, Sirohi, Chittorgarh, and Nagaur. These districts receive relatively high rain in short period in monsoon and remain dry in other months (Table 4.13).

An average high monthly rainfall is received by Chittorgarh followed by Rajkot, Bhavnagar, Surendranagar, and Udaipur. In June, Jaipur followed by Bhavnagar with 166.3 mm and 139.2 mm obtains highest monthly rainfall. Nevertheless, in July, Chittorgarh collects 286.6-mm-high rainfall in the region followed by Bhavnagar and Udaipur. Moreover, in August, Chittorgarh experiences high monthly rainfall with 286.6 mm, followed by Ajmer (189.7 mm) and Bhavnagar. In September, Surendranagar has 426.6-mm-high monthly rainfall followed by Rajkot and Bhavnagar having 370.2 mm and 279.6-mm-high monthly rainfall (Table 4.13).

4.6.2 North Deccan Zone: Land Degradation and Rainfall

In North Deccan region, 13 districts are opted for further analysis of rainfall. These districts are highly degraded in this zone. Pune, which is a highly degraded district in North Deccan region (according to Land Degradation Index) and possesses 18.54% (2900.09 km^2) areas under wasteland of the selected indicators of the total geographical area, receives low rainfall compared to some districts. Besides this, it also receives medium rainfall in the monsoon season. Pune receives rainfall of

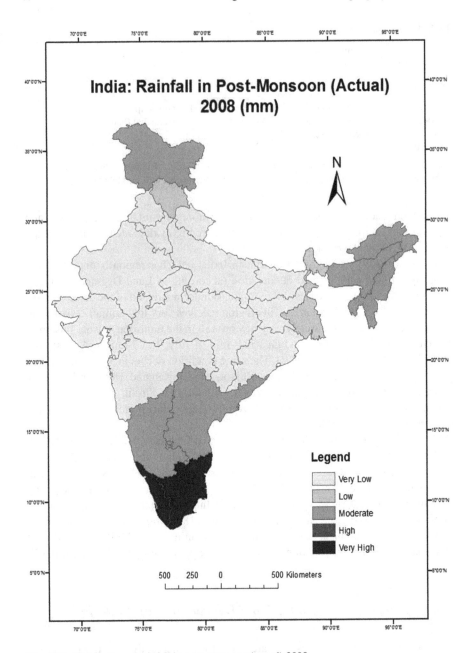

Fig. 4.12 Distribution of rainfall in post-monsoon (actual) 2008

Table 4.9 Average annual rainfall (normal) in India: post-monsoon season (2008)

Very high (>1500 mm)	Kerala, Karnataka, Assam, Meghalaya, Sikkim, West Bengal, Nagaland Manipur Mizoram, Tripura, Uttaranchal, Arunachal Pradesh
High (1200–1500 mm)	Jharkhand, Bihar, Maharashtra, Orissa, Himachal Pradesh, Jammu and Kashmir
Moderate (900–1200 mm)	Andhra Pradesh, Tamil Nadu, Uttar Pradesh, Madhya Pradesh
Low (600–900 mm)	Punjab, Gujarat
Very low (<600 mm)	Haryana, Chandigarh, Delhi, Rajasthan

148.4 mm in June, 99.5 mm in July, 216.1 mm in August, and 259.4 mm in September (Table 4.14). The same condition occurs in some districts of the region, namely, Satara and Parbhani. Some districts – Ahmadnagar, Nashik, Sangli, Sholapur, and Dhule – have obtained relatively low rainfall in 2008. Some districts – Raigarh, Ratnagiri, Kolhapur, Thane, and Sindhudurg – have received high monthly rainfall in the monsoon season in 2008 (Table 4.14).

Sindhudurg followed by Ratnagiri and Thane receives average high monthly rainfall. In June, highest monthly rainfall is obtained by Ratnagiri (871.7 mm) followed by Sindhudurg and Thane with 808.3 mm and 528 mm. However, in July, Thane collects 925-mm-high rainfall in the region followed by Ratnagiri and Sindhudurg. In August, Sindhudurg experiences high monthly rainfall with 898.5 mm, followed by Thane (826 mm) and Ratnagiri (898.5 mm). In September, again Sindhudurg has 645.1-mm-high monthly rainfall followed by Ratnagiri and then Thane having 645.1 mm and 520.9-mm-high monthly rainfall (Table 4.14).

4.6.3 South Deccan Zone: Land Degradation and Rainfall

In the North Deccan region, based on degraded land, 13 districts are culled out from this zone for assessing the rainfall properties. Cuddapah that is a highly degraded district (according to Land Degradation Index) and possesses 15.14% (2325.99 km²) areas under wasteland of the selected indicators of the total geographical area, has less than 33% vegetation cover. Besides this, it is also comparatively one of the rainfall districts. In 2008, Cuddapah receives rainfall of 8.9 mm in June, 45.1 mm in July, 97.9 mm in August, and 107.1 mm in September (Table 4.15).

Most of the districts receive very low rainfall less than 75 mm average monthly rainfall in 2008. Some districts – Warangal, Nellore, and Chandrapur – have received comparatively high average monthly rainfall in 2008 (Table 4.15). Average high monthly rainfall is received by Warangal followed by Nellore, Chandrapur, and Medak.

In June 2008, highest monthly rainfall is obtained by Warangal (190.1 mm) followed by Chandrapur and Karimnagar with 141.5 mm and 131.9 mm. While in July,

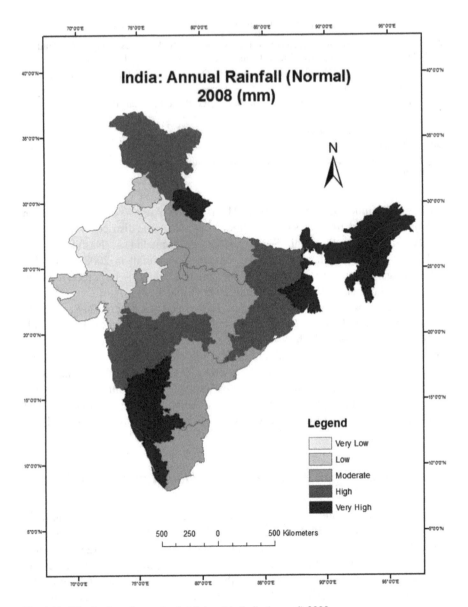

Fig. 4.13 Distribution of annual rainfall (mm) in India (normal) 2008

Chandrapur collects 318.2 mm which is high rainfall in the region followed by Warangal, Karimnagar, and Medak. In August, Warangal experiences high monthly rainfall again with 491.4 mm and then Chandrapur (398.6 mm) followed by Medak and Karimnagar (356.6 mm and 302.7 mm, respectively). In September, Anantapur has 193.7-mm-high monthly rainfall followed by Gulbarga, Warangal, Karimnagar, and Medak (Table 4.15).

Table 4.10 Average annual rainfall (actual) in India: post-monsoon season (2008)

Very high (>1500 mm)	Kerala, Karnataka, Assam, Meghalaya, Sikkim, West Bengal, Orissa
High (1200–1500 mm)	Jharkhand, Bihar, Nagaland Manipur Mizoram, Tripura, Maharashtra, Uttaranchal
Moderate (900–1200 mm)	Andhra Pradesh, Jammu and Kashmir, Arunachal Pradesh, Tamil Nadu, Himachal Pradesh, Uttar Pradesh
Low (600–900 mm)	Punjab, Haryana, Chandigarh, Delhi, Gujarat, Madhya Pradesh
Very low (<600 mm)	Rajasthan

4.6.4   Northeastern Zone: Land Degradation and Rainfall

In the northeastern zone, seven districts have been sorted for analysis. Karbi Anglong, one of highly degraded district (according to Land Degradation Index) and sixth rank in India, possesses 21.82% (2276.23 km^2) areas under wasteland of the selected indicators of the total geographical area. It has high vegetation cover (76.27% of total geographical areas) in 2008. Nevertheless, it is also comparatively one of the low monthly rainfall districts. In 2008, Karbi Anglong received rainfall of 164.1 mm in June, 169.4 mm in July, 266.5 mm in August, and 241.2 mm in September (Table 4.16).

Most of the districts receive very high rainfall more than 90 mm average monthly rainfall in 2008. Some districts – West Khasi Hills, West Kameng, and Upper Subansiri – have received comparatively very high average monthly rainfall with 573.77 mm, 271.39 mm, and 132.13 mm, respectively, in 2008. Average high monthly rainfall is received by Warangal followed by Nellore, Chandrapur, and Medak (Table 4.16).

In June 2008, highest monthly rainfall is obtained by West Khasi Hills (1211.4 mm) followed by West Kameng, Upper Subansiri, Senapati, and Ukhrul with 1103.4 mm, 288.4 mm, 275.8 mm, and 275.8 mm, respectively. Similarly, in July, West Khasi Hills collects 2035 mm which is high rainfall in the region followed by West Kameng, Upper Subansiri, and Tamenglong. In August, West Khasi Hills experiences high monthly rainfall again with 1899.8 mm, and then West Kameng (502.5 mm), followed by Upper Subansiri and Tamenglong (352.7 mm and 256 mm, respectively). In September, West Kameng has reregistered high monthly rainfall (591.5 mm) followed by Ukhrul, Senapati, and West Kameng (Table 4.16).

4.7   Land Degradation: Impact of Rainfall in Different Zone

From the above analysis, it is enunciated that extreme rainfall events become stronger in monsoon season, while the winter month and pre-monsoon and post-monsoon months receive very low rainfall. Seasonally high rainfall promotes vegetation density in the region, and vegetation cover is highly associated with rainfall (Paeth and

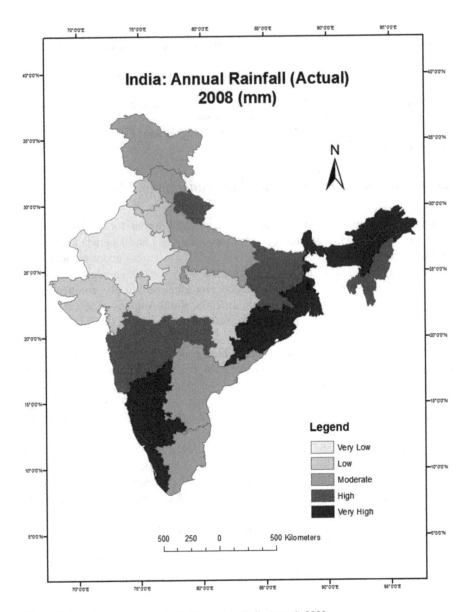

Fig. 4.14 Distribution of annual rainfall (mm) in India (actual) 2008

Thamm 2007). Conversely low rainfall is responsible for low vegetation growth in the region, and one study by Chauhan (2003) has suggested that desertification might be completely arrested by plantation.

In multilinear regression model, R represents as predictors of the dependent variables. Whereas R square represents the coefficient of determination, that is,

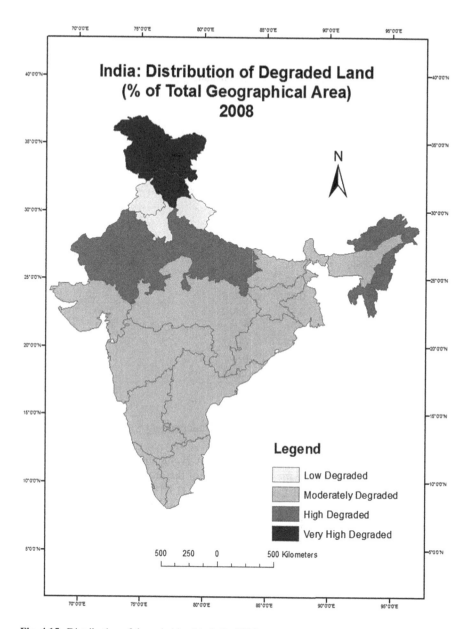

Fig. 4.15 Distribution of degraded land in India 2008

proportion of variation in the dependent variable which can be explained by the independent variable. However, adjusted R square is also needed to explain the accurate result reported by data. F-ratio tests whether the overall regression is a good fit for the data. Unstandardized coefficients indicate to what extent the dependent variable varies with an independent variable, when all other variables are held

Table 4.11 Land degradation and rainfall in degraded districts: model summary

R	R square	Adjusted R square	Std. error of the estimate	F value
0.80	0.65	0.53	461.47	5.47*

a. Predictors: (constant), rainfall of different month in mm
b. Dependent variable: LDI
*Significant at the level of <0.001

Table 4.12 Land degradation and rainfall in degraded districts: coefficients

	Unstandardized coefficients		Standardized coefficients		
	B	Std. error	Beta	t	Sig.
(Constant)	698.37	154.92	–	4.51	0.00
Jan	−4.45	8.86	−0.10	−0.50	0.62
Feb	11.01	6.38	0.30	1.72	0.09
Mar	−5.79	2.32	−0.47	−2.50	0.02
Apr	−0.46	4.61	−0.03	−0.10	0.92
May	−3.18	2.75	−0.33	−1.16	0.25
June	−1.07	0.84	−0.43	−1.28	0.21
July	1.06	0.73	0.50	1.46	0.15
Aug	0.11	0.94	0.05	0.11	0.91
Sep	−1.88	0.73	−0.42	−2.58	0.01
Oct	5.51	2.99	0.59	1.84	0.07
Nov	−5.51	1.92	−0.69	−2.88	0.01
Dec	111.92	15.74	0.81	7.11	0.00

constant. For the statistical significance of each independent variable, t-test has been run to test each independent variable.

R-value in above model represents the value of R, the multiple correlation coefficients, considered one of the measure qualities of the prediction of the dependent variables. In the model, Land Degradation Index (LDI) is a dependent variable. Average monthly rainfall (mm) in June, July, August, and September is an independent variable that is regressed on LDI. In Thar region, the regression model has been operated, R square, which is the coefficient of determination, and it explains 28.7% variability of the dependent variable[3]. F-ratio in Table 4.17 shows that F (4, 60) − 1.108, p < 0.401. It means that the regression model is fit at the level of 60% (Table 4.18).

In the North Deccan region, regression model has been tried out, where R = 0.340 represents about 34% correlations between the LDI and monsoon rainfall. R square, which is coefficient of determination, explains 0.12% variability of the dependent

---

[3] The degradation of land is not only influenced by rainfall in Thar region; in Chapter 2, vegetation is also one of the factors that cause degradation. High rainfall in a short period might cause soil degradation, but the study by Chauhan (2003) and Gaur and Gaur (2004) in the Thar region illustrated that vegetation cover is strongly helpful to promote the recovery of degraded land. It means that in Thar region, lack of vegetation is a major cause of succession of degraded land.

Table 4.13 District-wise monthly rainfall (mm) in Thar region

District	Jan	Feb	Mar	Apr	May	June	July	Aug	Sep	Oct	Nov	Dec
Jaisalmer	1.4	0.4	0	0.4	1.4	5.1	52.9	153.7	11.3	0	0	20.6
Kuchchh	0.1	0.1	0	0	0	63.2	38	84	113.2	0.1	1.1	14.6
Barmer	0	0	0.5	18.3	0	30	83.7	136.7	4.6	0	0	6.4
Rajkot	0	0.2	0	0	0	91.9	156.4	164.2	370.2	10.1	0.1	0.1
Pali	0	0	0	3.7	0	20.3	40.6	170.1	45.9	0	0	0
Rajsamand	0	0	0	0	0	34.4	60.9	142.1	77	0	0	0
Udaipur	0	0	0	0.5	5.9	60.6	211.4	178.1	128.9	8.7	0	0.4
Bhilwara	0	0	0	13.1	4.1	94.3	128.7	144.3	120.7	0	0.1	0
Ajmer	0	0	0.4	5.7	15.4	89.7	86.4	189.7	85.9	19.1	0	0
Jaipur	0	0	0	7	13	166.3	120.1	132.7	114.3	0.2	0	0
Surendranagar	0	0	0	2.7	0	37.2	90.6	164.6	426.6	8.1	0	8.7
Jamnagar	0.5	0.7	0.1	0	0	139.2	156.9	83.7	192.6	0.8	0.5	2.5
Bhavnagar	0	0	0	1.9	0	39.8	221.1	186.6	279.6	11.7	0	0
Sirohi	0	0	0.5	5.5	2	52.6	168.2	164.8	71.5	1.5	0.2	1.3
Chittorgarh	0	0	1.2	12	6.2	98.6	286.6	279.8	121.2	8.2	0	0
Nagaur	0	0	0	3	11.3	90	72.7	162.2	44	0	0	0

Source: IMD, GoI

Table 4.14 District-wise monthly rainfall (mm) in the North Deccan region

District	Jan	Feb	Mar	Apr	May	June	July	Aug	Sep	Oct	Nov	Dec
Thane	0	0	0	0	0	528	925	846	443.9	37.7	0.1	0.1
Nashik	0	0	0	1	0	53.2	246.2	253.7	475.5	53.8	0.2	0.5
Pune	0	0	13.7	0.8	0.8	148.4	99.5	216.1	259.4	60.1	4.4	6.8
Raigarh	22.4	21.3	27.8	31.5	2.3	256.7	298.1	586.5	261.5	5.8	0	0
Ahmadnagar	0	0	51.9	4.5	0	37.7	65.2	108.4	327.3	53	0.9	0
Satara	0	0	31.4	2.4	17.6	160.6	99.3	265.1	216	62.2	4.9	2.6
Ratnagiri	0	0	5.2	0	40.5	871.7	660.3	826.4	520.9	55.2	0.1	10.9
Kolhapur	0	0	56.5	3.3	69.1	435.8	329.6	671.3	419.1	91.8	1.8	3.7
Sindhudurg	0	0	47.3	0	0	808.3	619.8	898.5	645.1	32.6	0	7.3
Sangli	0	0	59.8	30.5	25.6	79.9	36	170.7	207.2	110.6	5.5	0.2
Kolhapur	0	0	56.5	3.3	69.1	435.8	329.6	671.3	419.1	91.8	1.8	3.7
Sholapur	0	0	49.7	7.7	4.8	27.7	63.4	156.2	279.1	58	25.4	0.5
Parbhani	3.5	0	7.1	12.1	0.5	110.4	152.6	128.9	239	19.7	35.3	0
Dhule	0	0	0	0	0	19.5	153.3	80	211	8.3	0	0

Source: IMD, GoI

variable, which is very bad explanation for the regression. F-ratio shows that F (4, 10) – 0.26, p < 0.894 (Table 4.19). It means that the regression model is not fit in this region. T-value for each month is also not significant, but at constant, it is significant at the level of 0.06 (Table 4.20). Some other factors might be present, which is associated with land degradation in the North Deccan region. One paper by Lal (2012) has estimated that in arid and semi-arid regions, it is very challenging to establish

Table 4.15 District-wise monthly rainfall (mm) in the South Deccan region

District	Jan	Feb	Mar	Apr	May	June	July	Aug	Sep	Oct	Nov	Dec
Medak	0	58.7	143.7	25.6	21	61.1	130.6	356.6	146.7	19.1	4.2	2
Mahbubnagar	0	7	124.5	14.9	8.5	49.9	76.8	185.7	109	26.2	19	0.4
Kurnool	0	12.8	99.5	0.1	34.5	42.2	100.7	137	129.3	90.1	61.3	1.1
Chittoor	15.9	3.4	51.7	3.1	78.5	37.6	99.6	82.7	146.6	170.3	253.1	10.8
Nellore	22.3	47	57.2	0	2.1	20.3	80.6	123.4	55.4	262.6	422.2	9.2
Prakasam	0	92.2	115	0.6	5.8	15.6	96.9	103.3	79.5	107.7	315.1	6
Cuddapah	0	33.7	47.6	0	26.7	8.9	45.1	97.9	107.1	95.3	185	1.4
Chandrapur	2.6	12.7	53.5	2.9	3.2	141.5	318.2	398.6	133.9	7.3	0	0
Anantapur	0	25.2	116.7	0	63	61.7	108	124.2	193.7	130.3	42.1	2.2
Gulbarga	0	4.5	143.1	6.1	5.1	62.2	102.9	152.6	168.7	74.4	16.3	8.1
Chitradurga	0	23.4	83.2	8.4	77.5	42.6	61.2	128.7	127.3	90.8	15	0.7
Karimnagar	0.2	1	98.7	5.4	8.1	131.9	184.2	302.7	150.3	11.5	3.6	0
Warangal	0	32.2	136	38.3	10.8	190.1	210.4	491.4	168.3	46.7	2.4	0

Source: IMD

cause and effect relation between soil degradation and climate at desired spatial and temporal scale. The Vidarbha region of this area has suffered due to mining practices and water erosion (Ghatol and Karale 2000).

In the South Deccan region, regression model has been operated, where R = 0.641 represents about 64% correlations between selected variables. R square explains 41.1% variability of the dependent variable. F-ratio (Table 4.21) explains that F (4, 70) – 1.40, p < 0.40. It means that the regression model is fit at the level of 60%. T-test is significant at constant, but not for all the indicators (Table 4.22).

Regression model has been run in the North Deccan region to know the measure quality of the dependent variable LDI of some selected districts, where R = 0.747 represents about 75% correlation between the dependent and independent variable. R square explicates 56% variability of the dependent variable. F-ratio distinguished that F (4, 32) – 1.108, p < 0.689. It means that the regression model is fit at the level of 32% (Table 4.23). While individual indicators do not significantly affect land degradation (Table 4.24).

Overall, it is concluded that rainfall is one of the major factors to determine the changing condition of land degradation. In the northeastern region, rainfall is highly correlated with LDI compared to other zones of degraded land.

4.8 Conclusion

As it is concluded, desertification is a slow process. Only rainfall is not major factor, but to some extent, it also affects soil erosion. It is observed that rainfalls received by the regions are high during summer monsoon, while arid and semi-arid region rainfall obtained in June, July, August, and September and the remaining months

Table 4.16 District-wise monthly rainfall (mm) in the northeastern region

District	Jan	Feb	Mar	Apr	May	June	July	Aug	Sep	Oct	Nov	Dec
West Khasi Hills	40.6	21.5	233.7	257	295.7	1211.4	2035	1899.8	591.5	292	2.7	4.3
Karbi Anglong	13.1	14.8	50.2	47	118.8	164.1	169.4	266.5	241.2	100.3	0	0
Ukhrul	0	5	29	30.8	106.2	275.8	14.2	329.4	279.6	19.2	0	12
Kameng West	40.9	26.8	45.4	79.6	357.3	1103.4	502.5	576.7	253.5	268	0.4	2.2
Subansiri Upper	79.8	12.8	140	141.8	83.7	288.4	352.7	217.6	120.6	146.6	0	1.6
Tamenglong	33.3	31.7	78.9	14.8	93.2	170.1	256	200.5	85.6	95.8	0.9	2
Senapati	0	5	29	30.8	106.2	275.8	14.2	329.4	279.6	19.2	0	12
Ukhrul	0	5	29	30.8	106.2	275.8	14.2	329.4	279.6	19.2	0	12

Source: IMD, GoI

Table 4.17 Land degradation and rainfall in Thar region: model summary

R	R square	Adjusted R square	F
0.536	0.287	0.028	1.11*

a. Predictors: (constant), rainfall in June, July, August, and September
b. Dependent variable: LDI
*Significant at the level of <0.40

Table 4.18 Land degradation and rainfall in Thar region: coefficient

Variable	Unstandardized coefficients		Standardized coefficients	t	Sig.
	B	Std. error	Beta		
Constant	2897.72	1296.38		2.24	0.047
June	−8.90	7.32	−0.36	−1.22	0.249
July	−0.79	5.78	−0.05	−0.14	0.89
August	−6.27	8.49	−0.26	−0.74	0.476
September	−1.95	2.42	−0.22	−0.80	0.438

Table 4.19 Land degradation and rainfall in the North Deccan region: model summary

R	R Square	Adjusted R Square	F
0.340	0.115	−0.33	0.26*

a. Predictors: (constant), rainfall in June, July, August, and September
b. Dependent variable: LDI
*Significant at the level of <0.89

Table 4.20 Land degradation and rainfall in the North Deccan region: coefficients

Variable	Unstandardized coefficients		Standardized coefficients	t	Sig.
	B	Std. error	Beta		
Constant	536.33	246.76		2.17	0.06
June	0.11	0.74	0.15	0.15	0.885
July	−0.41	0.62	−0.51	−0.65	0.533
August	0.49	0.82	0.69	0.60	0.566
September	−0.51	0.90	−0.32	−0.57	0.586

has highly dried up. The pattern of rainfall in different year shows an increasing trend, and degraded land reveals a decreasing trend. On the other hand, low-pressure area has increased during monsoon months. Consequently, many natural disturbances occur. Rainfall is determined by other factors. In arid and semi-arid regions, lack of rainfall is a causal factor to promote less vegetation cover. It is estimated in the previous chapter that increase in the unforested land causes high land degradation. The analysis brings out that in the northeastern region, increase in soil degradation is highly associated with rainfall because this region receives very high and intense rainfall.

Proceeding further, looking at the climatic impact, rainfall is a major factor that influences both land degradation and agricultural productivity. Therefore, it has been

Table 4.21 Land degradation and rainfall in the South Deccan region: model summary

R	R square	Adjusted R square	F
0.641	0.41097	0.11646	1.4*

a. Predictors: (constant), Rainfall in June, July, August, and September
b. Dependent variable: LDI
*Significant at the level of <0.40

Table 4.22 Land degradation and rainfall in the South Deccan region: coefficients

Model	Unstandardized coefficients		Standardized coefficients	t	Sig.
	B	Std. error	Beta		
Constant	527.11	265.55		1.98	0.082
June	−2.51	3.12	−0.66	−0.80	0.445
July	0.94	1.50	0.33	0.63	0.548
August	−0.46	1.02	−0.30	−0.45	0.664
September	0.33	1.90	0.06	0.17	0.867

Table 4.23 Land degradation and rainfall in the northeastern region: model summary

R	R square	Adjusted R square	F
0.747	0.55773	−0.3268101	0.63*

a. Predictors: (constant), rainfall in June, July, August, and September
b. Dependent variable: LDI
*Significant at the level of <0.689

Table 4.24 Land degradation and rainfall in the northeastern region: coefficients

Model	Unstandardized coefficients		Standardized coefficients	t	Sig.
	B	Std. error	Beta		
Constant	−9.269	392.345		−0.024	0.983
June	−0.131	0.407	−0.283	−0.323	0.777
July	1.431	1.392	4.815	1.030	0.411
August	−2.788	2.576	−8.073	−1.083	0.392
September	4.925	3.883	3.827	1.268	0.332

concluded that desertification is a retarded process, to which rainfall influences to some extent. Only rainfall is not major factor, but to some extent, it also affects soil erosion during heavy rainfall and also during lack of rainfall. It is observed that rainfalls received by the Indian subcontinent are high during summer monsoon, while arid and semi-arid region rainfall obtained in June, July, August, and September and the remaining months has dried up rapidly. Both factors responsible for land degradation, for instance, heavy and splash rainfall, cause high soil erosion, and lack of rainfall does not support the vegetation improvement and survival. Thus, as it is probed in the third chapter, the lack of vegetation is responsible for land degradation. The pattern of rainfall in different year shows an increasing trend, and degraded land reveals a decreasing trend. On the other hand, low-pressure area has increased during monsoon months. Consequently, lot of natural disturbances occur. In the northeastern region of India, increase in soil degradation is highly associated with rainfall because this region receives very high and intense rainfall.

References

Chauhan, S. S. (2003). Desertification Control and Management of Land Degradation in the Thar Desert of India. The Environmentalist, 23(3), 219–227. https://doi.org/10.1023/B:ENVR.0000017366.67642.79

Francis, P. A. & Gadgil, S. (2006). Intense rainfall events over the west coast of India. Meteorology and Atmospheric Physics, 94, 27–42. https://doi.org/10.1007/s00703-005-0167-2

Grepperud, S. (1997). Poverty, Land Degradation and Climatic Uncertainity. Oxford Economic Papers, New Series, 42 (4), 586–608. Retrieved from http://www.jstor.org/stable/2663694

Gaur, M. K. & Gaur, H. (2004). Combating Desertification: Building on Traditional Knowledge Systems of the Thar Desert Communities. Environmental Monitoring and Assessment, 99, 89–103.

Ghatol, S. G. & Karale, R. L. (2000). Assessment of the Degraded Lands of Vidarbha Region Using Remotely Sensed Data. Journal of the Indian society of Remote Sensing, 28 (2 & 3), 213–219.

Harden, C. P., & Mathews, L. (2000). Rainfall response of degraded soil following reforestation in the Copper Basin, Tennessee, USA. Environmental Management, 26(2), 163–174. https://doi.org/10.1007/s002670010079

India Meteorological Department (IMD). Government of India. http://www.imd.gov.in/section/hydro/distrainfall/districtrain.html

Jain, S. K., Keshri, R., Goswami, A. & Sarkar, A. (2010). Application of meteorological and vegetation indices for evaluation of drought impact: a case study for Rajasthan, India. Natural hazards, 54, 643–656. https://doi.org/10.1007/s11069-009-9493-x

Kishk, M. A. (1990). Conceptual Issues in Dealing with Land Degradation/Conservation Problems in Developing Countries. GeoJournal, 20 (3), 187–190. Retrieved from http://www.jstor.org/stable/41144633

Paeth, H. & Friederichs, P. (2004). Seasonality and time Scales in the Relationship between Global SST and African Rainfall. Climate Dynamics, 23, 815–837. https://doi.org/10.1007/s00382-004-0466-1

Paeth, H. & Thamm, H. (2007). Regional Modelling of Future African Climate North of 15° S Including Greenhouse Warming and Land Degradation. Climatic Change, 83, 401–427. https://doi.org/10.1007/s10584-006-9235-y

Lal, R. (2012). Climate Change and Soil Degradation Mitigation by Sustainable Management of Soils and Other Natural Resources. Agricultural Research, 1 (3), 199–212. https://doi.org/10.1007/s40003-012-0031-9

Meshesha, D. T., Tsunekawa, A., Tsubo, M., Ali, S. A., & Haregeweyn, N. (2014). Land-use change and its socio-environmental impact in Eastern Ethiopia's highland. Regional Environmental Change, 14(2), 757–768. https://doi.org/10.1007/s10113-013-0535-2

Milton, S. J., Dean, W. R. J., Plessis, M. A. D. & Siegfried, W. R. (1994). Conceptual Model of Arid Rangeland Degradation. BioScience, 44 (2), 70–76. Retrieved from http://www.jstor.org/stable/1312204

Meadows, M. E. & Hoffman, T. M. (2003). Land Degradation and Climate Change in South Africa. The Geographical Journal, 169, 168–167. Retrieved from http://www.jstor.org/stable/3451397

Murata, F., Hayashi, T., Matsumoto, J. & Asada, H. (2007). Rainfall on the Meghalaya plateau in northeastern India- one of the rainiest places in the world. Natural Hazards, 42, 391–399.

Qiang, Z., XiuWan, C., QiXiang, F., HePing, J. & JiRen, L. (2011). A New Procedure to Estimate the Rainfall Erosivity Factor based on Tropical Rainfall Measuring Mission (TRMM) Data. Science China Technological Sciences, 54 (9), 2437–2445. https://doi.org/10.1007/s11431-011-4468-z

Ratna, S. B. (2012). Summer Monsoon Rainfall Variability over Maharashtra, India. Pure Applied Geophysics, 169, 259–273. https://doi.org/10.1007/s00024-011-0276-4

Ramachandran, S. & Kedia, S. (2013). Aerosol, clouds and rainfall: inter-annual and regional variations over India. Climate Dynamics, 40, 1591–1610. https://doi.org/10.1007/s00382-012-1594-7

UNCCD (1994). United Nations Convention to Combat Desertification in Countries Experiencing serious drought and/or Desertification particularly in Africa. United Nations, New York.

Zhang, C., Wang, X., Li, J. & Hua, T. (2011). Role of Climate and Human Intervention in Land Degradation: A Case Study by Net Primary Productivity Analysis in China's Shiyanghe Basin. Environmental Earth Science, 64, 2183–2193. https://doi.org/10.1007/s12665-011-1046-4

Chapter 5
Land Degradation and Agricultural Productivity

Abstract Land degradation is significantly associated with low agricultural productivity, but use of fertilizer has increased the productivity in different districts. Therefore, it makes measurement of land degradation impact complex eventhough rainfall remain the prime determinent of the agricultural productivity in India. The availability of irrigation, use of conventional inputs, credits, and extension services, and encouraging the adaptation of mechanical and chemical technologies have also boosted up the productivity. In India, rice and wheat are major food grain and are therefore the focus in the study. The chapter has dealt with only land degradation and indicators of fertilizer (N, P, and K) in one section which has concluded that these are significantly correlated with productivity of rice and wheat. The second section of the chapter looks at the relation of extent of rainfall, aggregate fertilizer, and soil degradation with rice and wheat production (yield/hectares).

Keywords Land degradation · Wasteland · Agricultural productivity · Regression analysis · Rice · Wheat

5.1 Introduction

Agricultural practices contribute to the economy of the country. Increase in agricultural productivity kicks in the growth in three ways: to provide capital for economic growth, provide employment, and increase the purchasing power of the rural people (Christensen and Yee 1964; Braun and Gerber 2012). In India, rice and wheat are major food grain (Easter et al. 1977) and are therefore the focus in the study. The agricultural production receives much more attention from social scientists with economic concern. Introducing new varieties, along with the increased availability of fertilizers and irrigation, highly raises the production potential. The study by Dobbes and Foster (1972) have summarized that these inputs- high yielding seeds, fertilizers, tube wells etc. attracts to large former but the former with smallholdings is not getting high benefits. While India stands in the same condition that the production is not improving fast, it is slow still. The study analyses that only irrigation and fertilizer do not affect significantly. If it happens like that, then all areas do not vary where irrigation and fertilizer facilitation is very improved (Easter 1977). It means the natural productivity of land influences, to some extent, the agricultural output. Everyone believes that green revolution comes to minimize the world's population problem. There is need to think that whether green revolution is a fact or myth, consequences of an agricultural discover in population control are threatening. To fulfil the food requirement of the world's population demands, high use of chemicals, which is a type of pollutant, directly affects the human and specially children's health (Paddock 1970).

Ranade (1986) has stated that it was technology, which brings changes in rice production, which also helped in tapping the comparative advantage of the state like Punjab and Haryana. The local shift effects took place only after technological changes (Adams and Bumb 1979). Consequently, the contribution of the pure yield effect was more in pre- than post-green revolution. This study also has concluded that no green revolution takes place in rice production in Punjab and Haryana. The new varieties of grains are highly responsive to fertilizers (Posgate 1974; Ranade 1986). Chakravarti (1973) has worked on green revolution and stated that by using high yielding varieties, the production is increased but relatively high in wheat (Priya and Pani 2017).

5.2 Methodology and Database

To ascertain the implication of land degradation, labour, tractors, tube wells, irrigation, and fertilizers used for agricultural practices, the methods are as follows.

5.2.1 Statistical Methods

Regression model has fitted to capture the effect of land degradation on agricultural productivity. Firstly, in this study, Value output per ha. of 32 crops has been taken for the study of 281 districts of India, where value output is taken as dependent variable. In India, productivity also varies. For example, due to more use of fertilizer and availability of irrigation, productivity is high in Punjab and Haryana. For assessment of this problem, fertilizer data with land degradation will be used. In the study, independent variable is land degradation index (LDI), workers (no. of agricultural workers per GCA ha), fertilizer (N + P + K per ha), tractors (no. of tractors per ha), tube wells (no. of tube wells per ha), and irrigation (percentage of GCA under irrigation). In order to probe the impact of land degradation status and fertilizer on the level of socio-economic development, linear regression model has been used.

$$Y = \beta 0 + \beta 1 X1 + \beta 2 X2 + \beta 3 X3 + \beta 4 X4 + \beta 5 X5 + \beta 6 X6 + u$$

Here,

Y is the value output of 35 crops.

$X1$ is workers (no. of agricultural workers per GCA ha), $X2$ is fertilizer (N + P + K per ha), $X3$ is tractors (no. of tractors per ha), $X4$ is tube wells (no. of tube wells per ha), $X5$ is irrigation (percentage of GCA under irrigation in the district), and $X6$ is Land Degradation Index as an independent variable.

5.2.2   Source of Data

Land degradation and wasteland data are taken from NRSC, and Land Degradation Index is self-calculated. Agricultural data is collected from the Bhalla and Singh's Work on Agriculture. The data is used of Year 2005–2008.

5.3   Samples for the Study

All districts of the whole India is selected for the study, but by Bhalla value output of 35 crops is calculated. Those districts are selected for the study. To assess the second last objective of the study, 281 districts are selected for measuring the impact of land degradation on agriculture production, respectively.

For the further study, ten highly degraded districts have been taken for the assessment of impact of land degradation with the other additional indicator. The other factors are selected for study because the Indian agriculture is not only influenced by land quality, but climatic characteristics like rainfall and fertilizers aggregately affect the production.

5.4   India: Analysis of Relation Between Land Degradation and Agriculture[1,2]

5.4.1   India: Implication on Agricultural Productivity

Crops are major factors making food available to human beings; therefore, to evaluate the impact assessment of land degradation, the following regression has been calculated. The linear relationship is strong but not very high. The value of $R^2 = 0.297$ expresses that about 30%, total variation is explained. The F value (18.55), which is

---

[1] This chapter is majorly taken content from the Unpublished M.Phil. Dissertation Submitted to Jawaharlal Nehru University, New Delhi, by the author in 2014.

[2] Priya and Pani (2017). Land Degradation and Agricultural Productivity: A District Level Analysis, India. Journal of Rural Development, 36 (4), 557–568. NIRD&PR, Hyderabad. ISS 0970-3357.

Table 5.1 India: Implication agricultural productivity: model summary

R	R^2	Adjusted R^2	Std. error of the estimate	F
0.540	0.292	0.276	4.080	18.55*

[a]Predictors: (constant), LDI, no. of agriculture workers, N + K + P, no. of tractors, no. of tube wells, irrigation
[b]Dependent variable: Value output of 35 crops
*Significant at the level of <0.001

Table 5.2 India: implication agricultural productivity: coefficients

	Unstandardized coefficients		Standardized coefficients		
	B	Std. error	Beta	t	Sig.
(Constant)	7.228	0.623		11.597	0.000
LDI	−0.002	0.001	−0.117	−2.223	0.027
No. of agricultural workers per GCA ha	−1.126	0.285	−0.213	−3.956	0.000
N + P + K per ha	10.188	1.904	0.301	5.352	0.000
No. of tractors per ha	−59.373	17.624	−0.213	−3.369	0.001
No. of tube wells per ha	15.543	2.970	0.286	5.233	0.000
Irrigation percentage of GCA under irrigation in the district	0.036	0.011	0.220	3.233	0.001

a p-value, gives the result of hypothesis, and it implies to test whole model. In this model, it is significant at the level of 0.001 (Table 5.1). t-test has been done for the testing significance of each independent variable. In the following result, t-value for each independent variable is highly significant (Table 5.2). It is important to note that land degradation index has negative and significant impact on agricultural productivity, even after controlling the other factors.

Some studies in India on agriculture productivity supports that the quality of land and fertilizers affects the rice production in India. At present, the formers of Andhra Pradesh are facing the problem in the market of poor grain quality (Anitha and Rajyalakshmi 2012). In Andhra Pradesh Groundnut, cereal and millets with rice are cultivated in the upland areas of the basin where the soil thickness is less than 1 m and rain is the only source of water. Sugarcane and rice crops are cultivated using dug, dug-cum-bore, bore wells for irrigation in the tank, pond, command areas, and topographical low lands where the thickness of soil is more than 2 m (Rao et al. 1997).

In Rajasthan, the degradation has taken place due to moderate water erosion, wind erosion, and gypsum quarry (Adams and Bumb 1979). A study has tried to correlate the satellite data with the physicochemical characteristics, and it has revealed that in western Rajasthan the mean organic carbon is 31% less in slightly, 62.7% in moderately, and 68% less in severely degraded soil than non-degraded soil. Impact of types of land degradation such as wind erosion, water erosion, and salinization on fertility status of soil has revealed loss of potassium, phosphorus, and organic carbon (Raina et al. 2009). Due to high salinity and alkalinity, gullied and ravines, absence of better irrigation system and lack of agriculture techniques and poor know -how to use the appropriate fertilizers and pesticides for their crops (Easter et al. 1977; Adams and Bumb 1979; Drost et al. 1999), the rice production has been impacted. Conversely, some states like Chhattisgarh, Orissa Punjab, and Haryana are using a good technologies and high-yielding variety (HYV) seed, fertilizers, and pesticides for agricultural practices, which increase the short time fertility of the land is doing very well in production and increasing the production[3] (Dayal 1984; McGuirk and Mundlak 1992).

5.4.2   Agricultural Productivity in High Land Degraded Districts

5.4.2.1   Sampled Districts: Implication on Rice

As it is explained above, this section deals only with the highly land-degraded districts which come under the rice production (821,135 ha area). Cuddapah, Karbi Anglong, Bhilwara, Prakasam, Chittoor, Anantpur, Vizianagaram (Vijayanagaram), Rajsamand, Sagar, and Nellore are selected as a sample, which have gross 876,929 ha cropped areas under rice in 2006–2007 collectively. All the districts have more than 9% degraded land out of total geographical area during 2007–2008.

Regression statistics explains that the relation of rainfall, land degradation, and fertilizer with rice production is very strong, having $R^2 = 0.694$ correlation. The variance is explained through r^2 by 69%. The whole regression is significant at the level of 0.055 (Table 5.3).

Table 5.3 Rice in sampled districts: model summary

R	R^2	Adjusted R^2	Std. error of the estimate	F
.833[a]	0.694	0.541	0.70	4.53*

[a]Predictors: (constant), average rainfall, total fertilizer (KNP) (metric tonnes/hectare), Land Degradation Index
[b]Dependent variable: rice (yields tonnes per hectare)
*Significant at the level of <0.05

---

[3] In this study, the purpose is only to assess the impact of land degradation, N, P, and K as fertilizer indicator, agriculture worker, tractor, tube wells, and irrigation on agriculture production. This is why the research does not focus on the impact of other factors like technologies and characteristics of the soil (Adams and Bumb 1979; Dayal 1984; McGuirk and Mundlak 1992). There are might be other factors responsible to affect agriculture.

Table 5.4 Rice in sampled districts: coefficients

	Unstandardized coefficients		Standardized coefficients		
	B	Std. error	Beta	t	Sig.
(Constant)	0.799	1.538		0.52	0.622
Land degradation	−0.013	0.032	−0.119	−0.399	0.704
Fertilizer	10.368	3.29	0.768	3.151	0.02
Average annual rainfall	0.004	0.012	0.091	0.324	0.757

Table 5.5 Wheat in sampled district: model summary

R	R^2	Adjusted R^2	Std. error of the estimate	F
0.955	0.912	0.868	0.234	20.85

[a]Predictors: (constant), average rainfall, total fertilizer (KNP) (metric tonnes/hectare), Land Degradation Index
[b]Dependent variable: wheat (yields tonnes per hectare)
*Significant at the level of <0.001

The t-value for each variables (land degradation, fertilizer, and average annual rainfall) is 0.399, 3.151, and 0.324, and it is significant at the level of 0.704, 0.02, and 0.757, respectively (Table 5.4).

5.4.2.2 Sampled Districts: Implication on Wheat

Again, ten districts, namely, Jaisalmer, Udaipur, Cuddapah, Karbi Anglong, Bhilwara, Pali, Raigarh, Anantpur, Ajmer, and Rajsamand, are selected that highly degraded under the wheat production (167,501 ha area) (Agriculture Census). Land degradation (percentage of TGA), fertilizer (metric tonnes/hectare), and average annual rainfall (mm) are regressed on the wheat-producing district (yield tonnes/hectare), and R^2 is 0.912. Thus, the variability is explained by 91% (Table 5.5).

The total model is significant at the level of 0.001. The probability value for f-test is 20.85, and the degree of freedom is three (Table 5.5). The t-value for land degradation is 7.51, for fertilizer is 3.14, and for average annual rainfall is 4.05. The significance of t-value for land degradation, fertilizer, and average annual rainfall is 0.000, 0.020, and 0.006, respectively (Table 5.6).

5.5 Conclusion

It is evident through the analysis that not only man-made environment affects the high productivity, but also to some extent rainfall and soil degradation impacts on production of rice and wheat (yield/hectares) too. However, it is also drawn a conclusion that in some states the encroachment of land degradation on wheat and rice does not come into appearance due to use of fertilizer and availability of irrigation system.

Table 5.6 Wheat in sampled district: coefficients

	Unstandardized coefficients		Standardized coefficients		
	B	Std. error	Beta	t	Sig.
(Constant)	−0.5748	0.3964		−1.4500	0.1973
Land degradation	19.6327	2.6156	1.1447	7.5061	0.0003
Fertilizer	0.0090	0.0029	0.4823	3.1437	0.0200
Average annual rainfall	0.0428	0.0106	0.5337	4.0464	0.0068

For the wheat region in India, the new varieties are introduced. In Uttar Pradesh, Punjab, and Haryana, development of canal system has come with opportunity to improve the productivity of land. The same things are responsible for rice too. Other than that, increment in eastern market structure, improvement in connectivity, developmental efforts, new crop technology, use of supporting inputs, and improvement in social and private irrigation in eastern India are creditworthy to increment in rice production besides lagging with the land degradation.

Furthermore, land degradation is significantly associated with low agricultural productivity, but use of fertilizer has increased the productivity in different districts. Due to this complexity of relation, it is very difficult to measure of land degradation impact even rainfall also wallop the productivity. The availability of irrigation, use of conventional inputs, credits, and extension services, and encouraging the adaptation of mechanical and chemical technologies have also boosted up the productivity. Nevertheless, the chapter has dealt with only land degradation and indicators of fertilizer (N, P, and K) in one section which has concluded that these are significantly correlated with productivity of rice and wheat. The second section of the chapter looks at the relation of extent of rainfall, aggregate fertilizer, and soil degradation with rice and wheat production (yield/hectares).

References

Adams, J. & Bumb, B. (1979). Determinants of Agricultural Productivity in Rajasthan, India: The Impact of Inputs, Technology, and Context on Land Productivity. Economic Development and Cultural Change, 27 (4), 705–722. Retrieved from http://www.jstor.org/stable/1153566

Anitha, G. & Rajyalakshmi, P. (2012). Value added products with popular low grade rice varieties of Andhra Pradesh. Journal of Food Science Technology, Doi:https://doi.org/10.1007/s13197-012-0665-4

Braun, J. V. & Gerber, N. (2012). The Economics of Land and Soil Degradation- Toward an Assessment of the Cost of Inaction. In R. Lal et al. (Eds.), Recarbanization of the Biosphere: Ecosystems and the Global Carbon Cycle, (pp.493–516). Doi: https://doi.org/10.1007/978-94-007-4159-1_23

Christensen, R. P. & Yee, H. T. (1964). The Role of Agricultural Productivity in Economic Development. Journal of Farm Economics, 46 (5), 1051–1061. Retrieved from http://www.jstor.org/stable/1236681

Chakravarti, A. K. (1973). Green Revolution in India. Annals of the Association of American Geographer, 63 (3), 319–330. Retrieved from http://www.jstor.org/stable/2561997

Dayal, E. (1984). Agricultural Productivity in India: A Spatial Analysis. Annals of the Association of American Geographer, 74 (1), 98–123. Retrieved from http://www.jstor.org/stable/2562616

Dobbes and Foster (1972). Incentives to Invest in New Agricultural Inputs in North India. Economic Development and Cultural Change, 1972, vol. 21, issue 1, 101–17. https://doi.org/10.1086/450611

Drost, H., Mahaney, W. C., Bezada, M. & Kalm (1999). Measuring the Impact of Land Degradation on Agricultural Production: A Multi-Disciplinary Approach. Mountain Research and Development, 19 (1) 68–70. Retrieved from http://www.jstor.org/stable/3674115

Easter, K. W., Abel, M. E. & Norton, G. (1977). Regional Differences in Agricultural Productivity in Selected Areas of India. American Journal of Agricultural Economics, 59 (2), 257–265. Retrieved from http://www.jstor.org/stable/1240015

Easter, K. W. (1977). Improving Village Irrigation Systems: An Example from India. Land Economics, 53, (1), 56–66. Retrieved from http://www.jstor.org/stable/3146108

McGuirk, A. M. & Mundlak, Y. (1992). The Transition of Punjab Agriculture: A Choice of Technique Approach. American Journal of Agricultural Economics, 74 (1), 132–143. Retrieved from http://www.jstor.org/stable/1242997

Paddock, W. C. (1970). How Is the Green Revolution?, BioScience, 20 (16), 897–902. URL: https://www.jstor.org/stable/1295581

Priya and Pani (2017). Land Degradation and Agricultural Productivity: A District Level Analysis, India. Journal of Rural Development, 36 (4), 557–568. NIRD&PR, Hyderabad. ISS 0970-3357

Posgate, W.D. (1974). Fertilizers for India's Green Revolution: The Shaping of Government Policy. Asian Survey, 14 (8), 733–750. University of California Press. DOI: https://doi.org/10.2307/2642724 Retrieved from http://www.jstor.org/stable/2642724

Ranade, C. G. (1986). Growth of Productivity in Indian Agriculture: Some Unfinished Components of Dharm Narain's Work. Economic and Political Weekly, 21 (25/26), A75–A77+A79–A80. Retrieved from http://www.jstor.org/stable/4375826

Raina, P., Kumar, M. & Singh, M. (2009). Mapping of Soil Degradation Hazards by Remote Sensing in Hanumangarh District (Western Rajasthan). Journal of Indian Society of Remote Sensing, 37, 647–657.

Rao, Y. S., Reddy, T. V. K. & Nayudu, P. T. (1997). Groundwater quality in the Niva River basin, Chittoor district, Andhra Pradesh, India. Environmental Geology, 32 (1), 56–63.

Chapter 6
Land Degradation: Reclamation and Policy Adaptation

Abstract Land Degradation Index for all these factors has illustrated very clearly that it is the time to call a return to environmental (soil as a part of our environment and source of livelihood) security with the food security to the large population of India. This chapter has tried to summarize previous chapters and accordingly suggested some macro-level policies, which is needed to be adopted urgently to achieve land degradation neutrality in 2030 as targeted under both UNCCD and Sustainable Development Goals (SDG-15.3).

Keywords Land degradation · Wasteland · Policy implication · Macro-level approach and micro-level approach

6.1 Introduction

Land has been the most precious resource after air and water, which is under severe danger and suffering from all the major land degradation. Among all, the degradation of forest has been more important. Declining vegetation accelerates the pace of land degradation process. Soil salinity/alkalinity has emerged as a threat to the sustainability of the agricultural land as well as the food security especially in India case, where population is the second largest of the world. It involves greater economic costs than thoughts. The study into various regions of India has found out that the severity of soil salinity has been increasing due to many human-induced factors, e.g. unscientific agricultural practices, over-irrigation, and irrigation by saline water due to non-availability of fresh water, etc. People at the village level, which are affected by soil salinity, have the good knowledge of the fact, if not best aware, and also traditionally have knowledge to overcome or reclaim the salty land. The integrated approach of modifying the plant according to environment and modifying soil to the need of plant could be more helpful. Indian farmers should be encouraged to take the steps by following the footstep taken by the government for solving land degradation problems (Picture 6.1).

Spatial distribution of land degradation in India focusing on the cause related to natural and man-made factors has unfolded many findings that inform about the

Picture 6.1 *"Usar"*, soil in the village Madanapur (Kannauj, Uttar Pradesh) containing salt. It is demarcated into a red line in the photograph which is largely between two lands. Other than demarcated land is agricultural land, which is being reclaimed by the locals. (Photograph taken during the fieldwork by the author in December 2017)

land degradation condition in India, especially in the context of gullied/ravine, scrublands, salinity, and alkalinity. Gullied/ravine problem is centralized in the Chambal region, while high salinity is concentrated in the central part of Uttar Pradesh and some parts of Gujarat and Tamil Nadu. In addition, taking consideration of scrubland then, it is distributed in all over the peninsula region, while the westernmost and easternmost part of Rajasthan, middle part of Western Ghats, and middle northeast regions are highly degraded. Land Degradation Index for all these factors has illustrated very clearly that it is time to call a return to environmental (soil as a part of our environment and source of livelihood) security with the food security to the large population of India.

However, this has not focused on policy issues of governmental concern. Still the land degradation problem with deforestation and population explosion are all needed to be taken care of by government, whose practices are viewed as contributing significantly to land degradation in this study. The studies in previous chapters have supported the fact to some extent that different varying causes of land degradatio might have impact on agriculture significantly.

Regression estimation of land degradation with population and non-forested land has uncovered that in a developing country like India, where population pressure and people under poverty line are in large amount, needs more concern. With the more development of technology and investment, the conservation of forested land, which is lacking in India because in India many districts do not have at least 33%

forest cover of total geographical areas, must be a part of motivation of the policies by the government.

What is more in the study is to detect the association of land degradation with agricultural productivity specially the production of major food grains (wheat and rice) with the controlling factors nitrate (N), phosphorus (P), and potash (K) (metric tonnes per hectare). As it is known, high productivity takes place in two different situations. In West India, it is associated with good irrigation system, a high level of purchased inputs, and relatively large holdings. In Andhra Pradesh, Tamil Nadu, and some districts of the West Bengal, it has happened due to good irrigation, a high level of labour input, and varying levels of purchased inputs and small holding. Whereas if focus is put on low productivity, then it again happened due to dependency on climate, insufficient moisture, and high population pressure. In both cases, physical environments are good, but a man-made environment is more favourable; therefore, one of the man-made factors, fertilizers, is also regressed. Thus, it is concluded that fertilizers increase the productivity instantly, but it does not happen for a long term. Degradation is a slow process. The impact of land degradation on productivity is also context specific.

The present status of land degradation has presented a threat or a danger to the food security for future especially in India. The information related land degradation is not enough to access the impact directly. There is a need to collect data at the local level and regional level. Whatever information is available by today has indicated that the risk of land degradation is higher with low input and subsistence agriculture than with agricultural systems. Land resilience is crucial to food production and other issues related to environment, sustainability, and development.

6.2 Policy Implications

What should be done to reduce this insecurity is a big concern. Public understanding of the long-term effect of cropland conservation and topsoil loss on food price has decreased the fertility rate of soil. There are several ways to improve soil erosion and degradation to secure food for the people and also reduce poverty such as replacement of more productive crops in the place of less productive crops, use of amaranths and other neglected traditional crop for increasing fertility, providing best opportunity for increasing income with labour-intensive agricultural technologies, avoiding disturbance and destruction of natural resources or ecosystem, restoration of soil quality, identification of alternate crops, improvement in efficiency of inputs, improvement in water productivity and soil fertility, micronutrient availability, adaptation of no-till farming, and conservation of agriculture, climate, etc. New innovation is also suggested like using remote sensing of plant nutritional stresses for targeted interventions, applying zeolites and nano-enhanced fertilizers and delivery system, improving biological nitrogen fixation and mycorrhizal inoculation, and conserving and recycling (waste water) water using drip/sub-drip irrigation, among others.

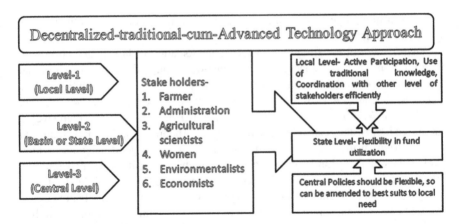

Fig. 6.1 Policy approach while dealing with soil salinity or broadly land degradation problem

Community-based soil water harvesting and soil conservation structure are playing a key role in improving surface and groundwater availability and controlling soil erosion in the watershed. Masonry check dam, low-cost earthen check dam, khadin system, farm ponds, gully check with loose boulder wall, intercropping system, mulching, strip cropping, afforestation, terrace farming and grassland management, and agroforestry are strategies for mitigation of land degradation (Fig. 6.1 and Pictures 6.2 and 6.3).

6.3   Macro-level Policy Suggestions

First and for the most important, the government should opt for an inclusive and comprehensive land degradation policy because having policy shows the attitudes, approaches, and priority level of the government. Secondly, the policy-making committee should not only include the highly qualified academicians and agriculturists but also include all the major and minor stakeholders like a representation of farmers, representation of those who implement the policy, ministries representation, etc. Features of policy related to land degradation at the national level (macro level) can be listed as follows:

1. The macro-level policy should have broad separate scientific objectives for the separate prominent types of land degradation, e.g. gully erosion, soil salinity, desertification and degradation of forest, and others which the government wants to prioritize for because each type of land degradation cannot be allotted the same finances. Priority of the types of land degradation should be based on the need for food security, environmental concerns, and resource utilization.
2. Regional objectives can be set because different regions have different types of land degradation problems. It will be good if decides according to the agro-

Picture 6.2 Terrace farming practices from various points in Darjeeling Region, West Bengal. (Taken by Author in 2020)

Picture 6.3 Image of farming under the cover of plastic to control soil moisture under commercial farming. Technically called mulching or plasticulture. (Photograph taken during the fieldwork at Firozpur, Punjab by the author in November 2017)

climatic region wise objectives for the reclamation of land degradation. This would help into the regionally balanced approach for solving the existing problems of land degradation.

3. Coordination among the ministries and among the states is of utmost importance for the implementation of land degradation policy due to various stakeholders' involvements. A separate body or Department of Land Resources can be declared as the nodal agency to track the efficient and effective performance of policy through coordination among the stakeholders.

4. The policy should be formulated at the national level in a way that it provides the scope to the states and districts and even further down the level of administration to change it according to the demand of their regional and local need.

5. Research and development into a different field of land degradation should be promoted especially in technological innovation sectors which have the solution approach to land degradation. At the same time, the traditional knowledge should be recognized with the best practices across the country. The special focus should be towards clubbing the technological advancement with the traditional knowledge to cope with land degradation problems.

6. More importantly, awareness and education programme for the farmers should be organized focusing towards the characteristics of different types of land degradation, implication on their land and socio-economic status of the farmers, best practices followed to reclamation or sustenance of land, best practices of sustainable farming, scientific approach of farming, etc.

These are the minimum policy-related recommendations at the national level, which can give a panoramic framework to be followed.

Index

Printed in the United States
by Baker & Taylor Publisher Services